Construction Digitalisation

This book explores construction digitalisation, particularly in developing countries. The book conceptualises a digitalisation capability maturity model that will enable construction organisations to self-assess and benchmark their digital capabilities in their quest for digital transformation.

Digitalisation offers a significant solution to the age-long problems of the construction industry. Research shows that when construction organisations transform from a traditional service delivery approach to a more digitalised approach, significant improvement in project delivery and better competitive advantage for these organisations will be attained. The attainment of these benefits is evident in developed countries where the digitalisation of construction activities continues apace. Unfortunately, the story is not the same for construction organisations in developing economies. While some organisations might be willing to be digitally transformed, most have no clue as how to go about it. To this end, this book provides guidelines for construction organisations seeking to transform their entities digitally. Its content is a valuable read for construction company owners as it provides a model which they can use in the digitalisation of their activities. Also, regulatory bodies in the construction industry can adopt the capabilities identified in the book as essential prerequisites for their members. Furthermore, the book serves as an excellent theoretical background reading for management researchers seeking to expand their knowledge on digitalisation of the construction industry and other associated industries.

Douglas Aghimien is a Postdoctoral Research Fellow at the cidb Centre of Excellence, Faculty of Engineering and the Built Environment, University of Johannesburg, South Africa.

Clinton Aigbavboa is the Director of the cidb Centre of Excellence and Sustainable Human Settlement and Construction Research Centre, University of Johannesburg, South Africa.

Ayodeji Oke is a Senior Lecturer at the Department of Quantity Surveying, School of Environmental Technology, Federal University of Technology Akure, Nigeria.

Wellington Thwala is the Chair of SARChI in Sustainable Construction Management and Leadership in the Built Environment, Faculty of Engineering and the Built Environment, University of Johannesburg, South Africa.

Routledge Research Collections for Construction in Developing Countries

Series Editors: Clinton Aigbavboa, Wellington Thwala, Chimay Anumba, David Edwards

A 21st Century Employability Skills Improvement Framework for the Construction Industry
John Aliu, Clinton Aigbavboa & Wellington Thwala

Construction Project Monitoring and Evaluation
An Integrated Approach
Callistus Tengan, Clinton Aigbavboa and Wellington Thwala

Developing the Competitive Advantage of Indigenous Construction Firms
Matthew Kwaw Somiah, Clinton Aigbavboa and Wellington Thwala

Construction Digitalisation
A Capability Maturity Model for Construction Organisations
Douglas Aghimien, Clinton Aigbavboa, Ayodeji Oke and Wellington Thwala

Sustainable Construction in the Era of the Fourth Industrial Revolution
Ayodeji E. Oke, Clinton O. Aigbavboa, Stephen S. Segun and Wellington D. Thwala

Construction Digitalisation

A Capability Maturity Model for
Construction Organisations

**Douglas Aghimien, Clinton Aigbavboa,
Ayodeji Oke and Wellington Thwala**

Routledge
Taylor & Francis Group

LONDON AND NEW YORK

First published 2021
by Routledge
2 Park Square, Milton Park, Abingdon, Oxon OX14 4RN

and by Routledge
605 Third Avenue, New York, NY 10158

Routledge is an imprint of the Taylor & Francis Group, an informa business

© 2021 Douglas Aghimien, Clinton Aigbavboa, Ayodeji Oke and Wellington Thwala

British Library Cataloguing-in-Publication Data
A catalogue record for this book is available from the British Library

Library of Congress Cataloging-in-Publication Data
A catalog record for this book has been requested

ISBN: 978-0-367-75854-7 (hbk)
ISBN: 978-0-367-75930-8 (pbk)
ISBN: 978-1-003-16465-4 (ebk)

Typeset in Goudy
by Apex CoVantage, LLC

This book is dedicated to God and the use and benefit of humanity

Contents

Preface

The construction industry is a crucial part of the socio-economic development of any country. The importance of the industry and the organisations therein in delivering adequate infrastructure in every nation has been well documented. In the same vein, the industry's failure in most developing economies to deliver projects within agreed project parameters and most importantly, to the specification of the client has also been adequately documented. This failure has, over time, been alluded to several issues, including poor technological uptake and advancement in the industry. With the fourth industrial revolution upon us, new, emerging, and disruptive technologies are changing the way human activities are being conducted. A crucial driver of this industrial revolution is the digitalisation of several human functions. Digitalisation offers solutions to the age-long problems of the construction industry. Equally, the attainment of these benefits is evident in developed countries where the digitalisation of construction activities has been successful. Unfortunately, the story is not the same for construction organisations in the developing economies. Albeit the apparent benefits of digitalisation to the industry, the embrace of this concept in most developing countries is still low. This situation is not unconnected with the numerous challenges facing construction industries within these countries. However, when construction organisations transform their traditional service delivery approach to a more digitalised method, significant improvement in project delivery and better competitive advantage for these organisations will be attained. To effectively adopt digitalisation concepts and infuse it into the already existing business model, a road map is needed. While some organisations might be willing to be digitally transformed, most have no clue as how to go about it. This situation begs the need for a capability maturity model for the digitalisation of construction organisations in developing economies.

This book, therefore, explores construction digitalisation, particularly in developing economies. The book conceptualises a digitalisation capability maturity model that will enable construction organisations to self-assess their digital capabilities in their digital transformation quest. Furthermore, the book promises value to readers as no existing study has focused on developing a digitalisation capability maturity model that can be used by construction organisations, particularly in developing countries where digitalisation is still at its infancy stage. Construction organisations can adopt the conceptualised model as a guide towards the digitalisation of their business activities. The book will also support construction regulatory bodies' agendas to adopt the defined capabilities as essential prerequisites for their members' transformation in this era of digital revolution. The book's content will also be beneficial to researchers seeking to expand the frontiers of knowledge on the construction industry's digitalisation. To this end, the authors confirm that the text utilised in this book reflects an original idea and, where necessary, materials that are of benefit to the book are appropriately cited and referenced.

Douglas AGHIMIEN
Clinton AIGBAVBOA
Ayodeji OKE
Wellington THWALA

Abbreviations

3D	Three dimensional
AEC	Architecture, Engineering and Construction
4IR	Fourth Industrial Revolution
AI	Artificial Intelligence
AR	Augmented Reality
BIM	Building Information Modelling
CAD	Computer-Aided Design
CIDB	Construction Industry Development Board
CMM	Capability Maturity Model
CMMI	Capability Maturity Model Integration
CSP	Communication Service Provider
DCMM	Digitalisation Capability Maturity Model
DCT	Dynamic Capability Theory
Df	Degree of Freedom
DL	Digital Library
DMMTP	Digital Maturity Model for Telecommunication Providers
DTMM	Digital Transformation Maturity Model
EIU	Economist Intelligence Unit
EPWP	Expanded Public Works Programme
FDM	Fused Deposition Modeling
FinTech	Financial Technology
GDP	Gross Domestic Product
H&S	Health and Safety
ICT	Information and Communication Technology
ILO	International Labour Organisation
IMD	Institute of Management Development
IoT	Internet of Things
IQD	Interquartile Deviation
IT	Information technology
OHS	Occupational Health and Safety
PMBOK	Project Management Body of Knowledge
R&D	Research and Development
RFID	Radio frequency identification technology

SEI	Software Engineering Institute
TMMM	Telemedicine Maturity Model
T-O-E	Technology-Organisation-Environment
UAV	Unmanned Aerial Vehicle
UK	United Kingdom
USA	United States of America
VR	Virtual Reality
X^2	Chi-square

Figures

Tables

Part I
Background information

Part I presents the overview of the book and likewise sets the foundation of what the book is all about.

1 General introduction

Introduction

The construction industry is one of the most critical sectors in the socio-economic development of any nation. The industry contributes significantly to infrastructure development, gross domestic product, direct and indirect employment in both developed and developing countries. Evidence of a healthy economy can be seen from the increase in construction activities through a series of infrastructure projects. Furthermore, the industry's physical product is crucial to the delivery of services of other sectors, as these other sectors of the economy require one form of physical structure or the other to perform their activities. Evidently, the construction industry's importance to the growth of both developed and developing countries cannot be overemphasised. However, the industry's impact is felt more in developing countries where the services of the industry are needed for the continuous provision of infrastructure that will help them measure up to their developed counterparts.

Despite the high importance attached to the construction industry, it is not without its challenges and difficulties. The industry is described as a demanding industry characterised by large, dynamic, and complex activities (Behm, 2008). These activities are evolving due to construction clients' growing demand, technological advancement resulting in the complex nature of designs, and rapid growth of innovations (Oke & Ogunsemi, 2011). Every day clients' taste and demand increases, and construction participants are saddled with the responsibility of meeting these demands through every method available. This creates a highly competitive environment, especially for construction organisations whose responsibility is to meet these clients' ever-changing tastes in return for financial gain. To stay relevant and survive in this competitive environment, these organisations have to continuously improve their service delivery by carefully planning their strategies and adopting innovative ideas (Aghimien et al., 2019b). To this end, imbibing the culture of "working smarter, not harder", has never been more important (Abidin et al., 2014; Dimick, 2014). This situation re-echoes the need for innovation in construction organisations. Similarly, since the construction environment is filled with uncertainties, organisations therein need some amount of contingency to improve their standard of service delivery, achieve quality, and attain optimum client satisfaction. Only through this can survival within the industry be assured.

The construction industry in most developing countries is failing to deliver successful projects. Evidence of projects being delivered above budget, beyond the agreed schedule, and below specification is recorded in these countries, leaving clients and other stakeholders dissatisfied (Agarwal *et al.*, 2016; Khalfan & Anumba, 2000; Ogunsemi & Jagboro, 2006; Oshodi *et al.*, 2017). This poor performance has been attributed to several challenges faced by the industry in different countries. One of these is the slow pace of technological advancement that has characterised most developing countries. This slow pace of technological advancement is evident in the poor digital technology adoption experienced in the construction industry (Aghimien *et al.*, 2019a; Alaloul *et al.*, 2020; Boton *et al.*, 2020; Kane *et al.*, 2015; Oesterreich & Teuteberg, 2016). The industry in the developing countries is slow to implement technological innovations, and as a result, has been seen to lack innovativeness. Old approaches to construction, such as tape measures and paper drawings are still familiar sights within these industries (Hudson, 2017). Therefore, there is need for some cultural shift as sticking to traditional methods will only leave construction organisations at the tail of the value chain and create more dissatisfaction in the product they deliver.

No doubt, the fourth industrial revolution (4IR), also known as Industry 4.0 is upon us, and it is disrupting the way we live and function as individuals and as societies. With digitalisation being a core driver of 4IR (Bienhaus & Haddud, 2018; Li *et al.*, 2017; Schwab, 2017), construction organisations stand the chance of benefitting from this industrial revolution by adopting the digital technologies therein. Furthermore, with the clamour for the adoption of "Construction 4.0" (Boton *et al.*, 2020; Forcael *et al.*, 2020), coined from Industry 4.0, where ubiquitous technologies are used for real-time decision-making and construction projects are handled using digital technologies such as the Internet of Things (IoT), big data, artificial intelligence, additive manufacturing, virtual and augmented reality, new business models, and e-value chains (Bibby & Dehe, 2018; Oesterreich & Teuteberg, 2016), construction organisations in developing countries can digitally transform their service delivery and change the narrative of poor project delivery that has characterised their industry.

While the benefits of adopting digital technologies in the construction industry are enormous, one question remains pertinent; how matured are organisations in the construction industry of developing countries to fully and effectively adopt digitalisation in all aspects of their construction service delivery? It has been observed that digitalisation is not a one-off event but rather should be considered as a piecemeal activity (Strukova & Liska, 2012). Drastic digital changes within an organisation might not bring the digital transformation being sought but rather might lead to some negative outcomes such as agitation among workers due to fear of replacement. Organisations must understand that skills should be enhanced through digital technology adoption and not replaced by it. Therefore, knowing what skill to improve, what technology to adopt in enhancing this skill, and what process and strategy to take in achieving this skill improvement and at the same time achieve the organisational objective of digital transformation and better project delivery is essential. The implication of this is that digitalisation of construction organisations can be a complex task that requires proper planning and a clear road map. With the complex and dynamic nature of both the construction industry and the digital world (Navon, 2005; Solis, 2016), having a digitalisation road map that tells the digital maturity level of construction organisations and the direction towards achieving digital maturity is important.

It is pertinent to note that while several maturity models have been developed in different industries and domains to measure different situations, there is the absence of a maturity model designed to address digitalisation of the construction industry specifically. Herein lies the problem of this study. While several maturity models have been developed over time in different industries (De Bruin *et al.*, 2005), none has been targeted at the construction industry's digitalisation. The available digital maturity models have looked at technological issues, strategies, process and operations, organisational culture, and policy issues within the business, education, manufacturing, and telecommunication domains. Although the construction industry operates in a similar manner with these industries as it is business-oriented and delivers finished products to its consumers, some other variables are deemed pertinent to the industry and must be considered. It is against this theoretical backdrop that this book seeks to provide possible direction for attaining digital transformation of construction organisations by developing a conceptualised digitalisation capability maturity model (DCMM).

Digitalisation at a glance

The word "digitalisation" has come with diverse meaning since the advent of 4IR, which brings about using a more digitalised form of production. Digitalisation is now a common feature in individual devices and large industrial systems and has become a popular concept around the world today due to its ability to create efficiency in operations, effectiveness, and provide new opportunities (Dall'Omo, 2017). The term "digitalisation" is the increased use of digital or computer technology in an organisation's operations (Kalavendi, 2017). When organisations decide to be digitalised, they abandon the old analogue ways for a more digital-oriented approach.

In past studies, words like digitisation, digital technology, digital collaboration, and computerisation have been used in place of digitalisation. In some cases, this exchange has been made wrongly. Rouse (2017) viewed "digitisation" as the process of transforming information into digital format. By doing this, information is organised into distinct units of data that can be separately addressed. This is the binary data that computers and many devices with computing capacity can process. This implies that digitisation involves converting or representing something non-digital into a digital format, which can be used by a computing system for diverse purposes. Digitalisation, on the other hand, is the strategic realignment of investment with technology, business models, and processes to compete in an ever-changing digital world (Solis, 2016). On more simplified terms, Ochs and

Riemann (2018) described the term "digitalisation" as the integration of digital technologies into everyday life, through digitising anything capable of being digitised. These digital technologies are known to enable a large amount of information to be compressed on small storage devices for easy preservation and transportation. They include all types of electronic equipment and applications that use information in the form of numeric codes. These devices can be mobile phones, video cameras, tablets, laptops, computers, the internet, communication satellites, and other innovations geared towards making certain activities easier and more effective (Anjos-Santos *et al.*, 2016; Pullen, 2009). Digital technologies can also be viewed from three facets, and these are software, information technology (IT) equipment (computers and related hardware), and communications equipment (Dimick, 2014). What makes these technologies different from other earlier technologies is their ability to be re-programmed, the use of homogenous data, and their self-referential nature (Yoo *et al.*, 2010). Thus, while digitisation is seen as the digital transformation of specific processes that were not digital, using digital technologies, digitalisation is the deployment of these transformed processes into the organisation's business life.

Based on the aforementioned descriptions, in the context of this book, the term "digitalisation" is conceptualised as *the innovative use of digital technologies in the delivery of tangible and intangible services within a construction organisation to gain a competitive advantage over other competitors while providing better service delivery.* This implies that an organisation within the construction industry can be said to have attained digitalisation when it can optimally utilise existing digital technologies in the delivery of its services and at the same time gain better advantage over its counterparts in the industry. These digital technologies include Building Information Modelling (BIM), augmented and virtual reality, IoT, big data, robotics and automation, cloud computing, 3D printing, machine learning, among others. Viewing digitalisation from this perspective is premised on the fact that these different technological innovations are based on digital operations. Gerbert *et al.* (2016) mentioned that with digitalisation, construction around the world would soon be depicted by the connection of systems using sensors, intelligent machines, mobile devices, and new software applications that will all be integrated on a central platform BIM. This further strengthens the notion of this study that digitalisation can be regarded as the digitisation of construction processes using the aforementioned digital technologies to attain a digitally transformed built environment.

Objectives of the book

The concept of digitalisation is not entirely new to the construction industry in developing countries as evidence of the use of some digital technologies like BIM and cloud computing has been observed in past studies (Aghimien *et al.*, 2019a; Dos Santos *et al.*, 2015; Ibem & Laryea, 2014). While lack of awareness and required skills have been attributed to the slow adoption of some of these digital technologies (Boton *et al.*, 2020; Oke *et al.*, 2018), the absence of a road map for organisations that are willing to be digitally transformed can also be a significant barrier. It is not unlikely that some organisations might have the idea of digital transformation in the quest for better service delivery and improved competitive advantage but find it challenging to understand the right process to follow. The absence of a digital transformation road map tailored to suit the construction industry that will guide these organisations in their digitalisation journey reveals a crucial gap that needs to be filled. Therefore, the objective of this book is to conceptualise a capability maturity model that will help assess and to improve the digitalisation of construction organisations by unearthing the key capability areas wherein maturity needs to be attained for construction digitalisation to be achieved.

Value of the book

Past research studies seeking to improve project delivery in developing countries have been tailored towards the performance of construction projects, factors responsible, and several suggestions have been proffered (Hussin *et al.*, 2013; Ibrahim *et al.*, 2010; Kaming *et al.*, 1997; Memon *et al.*, 2010; Ogunsemi & Jagboro, 2006; Omoregie & Radford, 2006; Windapo & Cattell, 2013). However, as a result of 4IR, recent trends have shown that if construction project delivery is to improve, and construction companies are to gain better reputation, then the construction process must be rooted in digital knowledge and strategically driven by digital technologies. The strategic application of digital tools and the benefits to be derived at the design, construction, and operation phases of construction projects have been well documented (Agarwal *et al.*, 2016; Castagnino *et al.*, 2016; Delgado *et al.*, 2017; Gerbert *et al.*, 2016; Ibem & Laryea, 2016; Ikuabe *et al.*, 2020). However, there still exists a gap in the body of knowledge regarding the capabilities needed by construction organisations in their activities' digitalisation. No clear road map exists as to what these organisations need to assess to determine their digital capability level. Knowing "where we are" and "where we ought to be" is necessary to improve the industry. Drawing from the manufacturing industry where Capgemini Consulting (2014) suggested that companies should undertake a detailed digital maturity assessment to have a transparent view of their current level of digital readiness, it is equally essential for organisations in the construction industry to know their current level of digital maturity and how to prepare for future digital growth. Similarly, if construction organisations are to achieve headway in the quest to increase their service delivery capabilities, undertaking a digital maturity assessment that will serve as a road map towards achieving this quest is necessary. This study provides that road map needed for construction organisation seeking digital transformation.

Practically, the book's contribution is seen as the unearthing of the different capability areas that organisations need to give adequate attention in their quest for a more digitalised construction industry. Unlike the past, where construction organisations implement without guidance, the conceptualised DCMM can now serve as a road map towards attaining digital transformation. Construction organisations can utilise the developed model to assess their current digital capability maturity and understand where they need to improve. Similarly, through proper self-assessment and benchmarking of their digitalisation capabilities with other organisations within and outside the industry, improved digitalisation of the construction industry can be attained. This is because the conceptualised model recognises the importance of digitalising vital areas of the construction process. If this is done correctly, better service delivery can be attained from the industry, and a more digitalised construction industry can be seen in the long run. In the same vein, professional regulatory bodies, saddled with the responsibility of promoting and developing the construction industry delivery capability for social and economic growth can use the conceptualised model as an assessment tool for the digital adoption of construction organisations in the course of project delivery. Theoretically, this book serves as an excellent platform for future studies seeking to explore construction digitalisation and how maturity can be attained within construction organisations. The conceptualised DCMM contributes significantly to the existing body of knowledge on maturity models, as there is none targeting construction digitalisation.

Structure of the book

The book is divided into five parts with ten chapters tailored towards clarifying the concept of construction digitalisation and the capability maturity required for digital transformation. The first part begins with Chapter 1, which gives background information about the book. Part II of the book consists of three chapters designed to provide clarity to construction digitalisation. Chapter 2 explores the construction industry's features in developing countries with a view to pinpoint the need for digital transformation. The challenges deterring the industry's development and the industry's performance in line with specific success criteria are assessed. Chapter 3 illuminates the concept of 4IR and digitalisation by exploring 4IR in different sectors, including construction. The chapter also assesses digitalisation as a key driver of this industrial revolution. Chapter 4 is designed to give meaning to the concept of construction digitalisation. The digital technologies needed to attain construction digitalisation, the drivers, barriers, risks of construction digitalisation, and the impact of digitalisation on construction performance are discussed. Part III of the book focuses on the theoretical dimension of digitalisation capability maturity and consists of three chapters. Chapter 5 sheds light on the need for a capability maturity model by explaining the constituents and characteristics of a maturity model. Chapter 6 evaluates relevant maturity models with a view to identifying the capability areas needed for digitalisation. Chapter 7 explores the dynamic capability theory, which underpins the conceptualised DCMM. Part IV of the book focuses on the conceptual perspective of construction digitalisation capability maturity and consists of two chapters. Chapter 8 focuses on the model's conceptual dimension with all capability areas, along with their sub-attributes. Chapter 9 explores the conceptualised model in a developing country using expert opinion. The significance of the capability areas and their sub-attributes are assessed using a developing country's case study.

Summary

This chapter introduced the readers to the idea behind the conception of this book. Emphasis was placed on the poor performance of construction industries in developing countries worldwide and the urgent need for construction digitalisation to provide better service delivery and a digitally transformed construction industry. The chapter gave an overview of the concept of digitalisation. It also revealed a gap in existing knowledge on maturity models as none exists for assessing construction organisations' digitalisation capability maturity. Therefore, the study's objective was to fill this existing gap in knowledge by developing a DCMM that will serve as an assessment tool and a road map for construction organisations to be digitalised. The next chapter gives an in-depth description of the construction industry in developing countries, and this sets the base for digitalisation in the industry.

References

Abidin, N.Z., Adros, N.A. and Hassan, H. (2014). Competitive strategy and performance of quantity surveying firms in Malaysia. *Journal of Construction in Developing Countries*, 19(2): 15–32

Agarwal, R., Chandrasekaran, S. and Sridhar, M. (2016). *Imagining Construction's Digital Future*. Capital Project and Infrastructure, McKinsey and Company. Available at: www.mckinsey.com/industries/capital-projects-and-infrastructure/ourinsights/imagining-constructions-digital-future [accessed 19–11–2020]

Aghimien, D.O., Aigbavboa, C.O. and Matabane, K. (2019a). Impediments of the fourth industrial revolution in the South African construction industry. *Construction in the 21st Century, 11th International Conference (CITC-11)*, London, UK, 9–11 September, pp. 318–324

Aghimien, D.O., Aigbavboa, C.O. and Oke, A.E. (2019b). Digitalisation of construction organisations in South Africa: A dynamic capability theory approach (2019). Advances in ICT in design, construction and management in architecture, engineering, construction and operations (AECO). *Proceedings of the 36th CIB W78 2019 Conference*, North Umbria University at Newcastle, London, UK, 18–20 September, pp. 74–83

Alaloul, W.S., Liew, M.S., Zawawi, N.A.W. and Kenndey, I.B. (2020). Industrial revolution 4.0 in the construction industry: Challenges and opportunities for stakeholders. *Ain Shams Engineering Journal*, 11(1): 225–230

Anjos-Santos, L.M., El Kadri, M.S., Gamero, R. and Gimenez, T. (2016). Developing English language teachers' professional capacities through digital and media literacies: A Brazilian perspective. In Manzoor, A. (Ed.), *Editions of Handbook of Research on Media Literacy in the Digital Age*. IGI Publishers, Harrisburg, PA

Behm, M. (2008). The construction sector; Rapporteur's report; East Carolina University. *Journal of Safety Research*, 39: 175–178

Bibby, L. and Dehe, B. (2018). Defining and assessing industry 4.0 maturity levels – case of the defence sector. *Production Planning and Control*, 29(12): 1030–1043

Bienhaus, F. and Haddud, A. (2018). Procurement 4.0: Factors influencing the digitisation of procurement and supply chains. *Business Process Management Journal*, 24(4): 965–984

Boton, C., Rivest, L., Ghnaya, O., *et al.* (2020). What is at the root of construction 4.0: A systematic review of the recent research effort. *Archives of Computational Methods in Engineering*: 1–20

Capgemini Consulting. (2014). Digitising manufacturing: Ready, set, go! Available at: www.de.capgemini-consulting.com/resource-fileaccess/resource/pdf/digitizing-manufacturing_0.pdf [accessed 19–11–2020]

Castagnino, S., Rothballer, C. and Gerbert, P. (2016). *What's the Future of the Construction Industry?* World Economic Forum. Available at: www.weforum.org/agenda/2016/04/building-in-the-fourth-industrial-revolution/ [accessed 19–11–2020]

Dall'Omo, S. (2017). Driving African development through smarter technology. *African Digitalisation Maturity Report*: 1–45

De Bruin, T., Rosemann, M., Freeze, R. and Kulkarni, U. (2005). Understanding the main phases of developing a maturity assessment model. *Proceedings of the Australasian Conference on Information Systems (ACIS)*, Sydney, 30 November–2 December

Delgado, J.M.D., Oyedele, L., Akinade, O., Bilal, M., Akanbi, L. and Ajayi, A. (2017). BIM data model requirements for structural monitoring of built assets. *Proceedings of Environmental Design and Management International Conference (EDMIC)*, Held Between, 22–24 of May, at the Obafemi Awolowo University, Ile-Ife, Nigeria, pp. 118–126

Dimick, S. (2014). *Adopting Digital Technologies: The Path for SMEs*. The Conference Board of Canada, Ottawa, ON, pp. 1–13

Dos Santos, S.D., Vendrametto, O., González, M.L. and Correia, C.F. (2015). Profile of building information modeling – BIM – tools maturity in Brazilian civil construction scenery. In Umeda, S., Nakano, M., Mizuyama, H., Hibino, N., Kiritsis, D. and von Cieminski, G. (Eds.), *Advances in Production Management Systems: Innovative Production Management Towards Sustainable Growth*. IFIP Advances in Information and Communication Technology, Springer, Cham, pp. 291–298

Forcael, E., Ferrari, I., Opazo-Vega, A. and Pulido-Arcas, J.A. (2020). Construction 4.0: A literature review. *Sustainability*, 12: 1–28

Gerbert, P., Castagnino, S., Rothballer, C., Renz, A. and Filitz, R. (2016). *Digital Engineering and Construction: The Transformative Power of Building Information Modelling*. The Bolton Consulting Group, Boston, MA

Hudson, V. (2017). *The Digital Future of the Infrastructure Industry: Innovation 2050*. Balfour Beatty, London

Hussin, J.M., Rahman, I.A. and Memon, A.H. (2013). The way forward in sustainable construction: Issues and challenges. *International Journal of Advances in Applied Sciences*, 2(1): 15–24

Ibem, E.O. and Laryea, S. (2014). Survey of digital technologies in procurement of construction projects. *Automation in Construction*, 46: 11–21

Ibem, E.O. and Laryea, S. (2016). E-tendering in the South African construction industry. *Journal of Construction Management*, 17(4): 310–328

Ibrahim, R., Roy, M.H., Ahmed, Z.U. and Imtiaz, G. (2010). Analysing the dynamics of the global construction industry: Past, present and future. *Benchmarking: An International Journal*, 17(2): 232–252

Ikuabe, M., Aghimien, D.O., Aigbavboa, C.O. and Oke, A.E. (2020). Exploring the adoption of digital technology at the different phases of construction projects in South Africa. *10th International Conference on Industrial Engineering and Operations Management*, United Arab Emirates, Dubai, 10–12 March, pp. 1553–1561

Kalavendi, R. (2017). Digitisation and digitalisation: The two-letter difference. Available at: https://medium.com/@ravisierraatlantic/digitization-digitalization-the-2-letter-dif ference-59b747d42ade [accessed 19–11–2020]

Kaming, P.F., Olomolaiye, P.O., Holt, G.D. and Harris, F.C. (1997). Factors influencing construction time and cost overruns on high – rise projects in Indonesia. *Construction Management and Economics*, 15(1): 83–94

Kane, G.C., Palmer, D., Phillips, A.N., Kiron, D. and Buckley, N. (2015). *Strategy, Not Technology, Drives Digital Transformation*. MIT Sloan Management Review and Deloitte University Press, Westlake, TX

Khalfan, M.M.A. and Anumba, C.J. (2000). Implementation of concurrent engineering in construction – readiness assessment. *Proceedings of Construction Information Technology (CIT2000) Conference*, Reykjavik, Iceland, 28–30 June, Vol. 1, pp. 544–555

Li, G., Hou, Y. and Wu, A. (2017). Fourth industrial revolution: Technological drivers, impacts and coping methods. *Chinese Geographical Science*, 27(4): 626–637

Memon, A.H., Rahman, I.A., Abdullah, M.R. and Azis, A.A. (2010). Factors affecting construction cost performance in project management projects: Case of mara large projects. *Proceedings of Post Graduate Seminar on Engineering, Technology and Social Science*, University Tun Hussein Onn Malaysia, Malaysia

Navon, R. (2005). Automated project performance control of construction projects. *Automation in Construction*, 14: 467–476

Ochs, T. and Riemann, U.T. (2018). IT strategy follows digitalisation. In *Encyclopedia of Information Science and Technology*, 4th edition IGI Publishers, Hershey, PA

Oesterreich, T.D. and Teuteberg, F. (2016). Understanding the implications of digitisation and automation in the context of Industry 4.0: A triangulation approach and elements of a research agenda for the construction industry. *Computers & Industrial*, 83: 121–139

Ogunsemi, D.R. and Jagboro, G.O. (2006). Time-cost model for building projects in Nigeria *Construction Management and Economics*, 24: 253–258

Oke, A.E., Aghimien, D.O., Aigbavboa, C.O. and Koloko, N. (2018). Challenges of digital collaboration in the South African construction industry. *Proceedings of the International Conference on Industrial Engineering and Operations Management*, Bandung, Indonesia, 6–8 March, pp. 2472–2482

Oke, A.E. and Ogunsemi, D.R. (2011). Value management in the Nigerian construction industry: Militating factors and perceived benefits. *Proceeding of the second International Conference on Advances in Engineering and Technology*, Faculty of Technology, Makerere University, Uganda, 30 January–1 February, pp. 353–359

Omoregie, A. and Radford, D. (2006). Infrastructure delays and cost escalation: Causes and effects in Nigeria. *Proceeding of Sixth International Postgraduate Research Conference*, Delft University of Technology and TNO, Netherlands, pp. 79–93

Oshodi, O.S., Ejohwomu, O.A., Famakin, I.O. and Cortez, O. (2017). Comparing univariate techniques for tender price index forecasting: Box-Jenkins and neural network model. *Construction Economics and Building*, 17(3): 109–123

Pullen, D.L. (2009). Back to basics: Electronic collaboration in education sector. In Salmons, J. and Wilson, L. (Eds.), *Editions Handbook of Research on Electronic Collaboration and Organizational Synergy*. IGI Publishers, Harrisburg, PA

Rouse, M. (2017). Meaning of digitalization. Available at: http://whatis.techtarget.com/definition/digitization [accessed 19–11–2019]

Schwab, K. (2017). *The Fourth Industrial Revolution*, 1st edition. Crown Business, New York

Solis, B. (2016). *The Six Stages of Digital Transformation Maturity*. Altimeter Cognizant. Available at: www.prophet.com/2016/04/the-six-stages-of-digital-transformation/#:~:text=To

%20make%20this%20more%20actionable,%2C%20Technology%20Integration%2C%20Digital%20Literacy [accessed 6–8–2020]

Strukova, Z. and Liska, M. (2012). Application of automation and robotics in construction work execution. *AD ALTA: Journal of Interdisciplinary Research*, 2(2): 121–125

Windapo, A.O. and Cattell, K. (2013). The South African construction industry: Perceptions of key challenges facing its performance, development and growth. *Journal of Construction in Developing Countries*, 18(2): 65–79

Yoo, Y., Henfridsson, O. and Lyytinen, K. (2010). The new organizing logic of digital innovation: An agenda for information systems research. *Information Systems Research*, 21(4): 724–735

Part II

Digitalisation in the construction industry

Part II of this book focuses on construction digitalisation as it relates to the construction industry in developing countries. This part is divided into three chapters, Chapter 2 gives an understanding of the construction industry in developing countries. Chapter 3 illuminates the fourth industrial revolution and digitalisation concept, while Chapter 4 explores digitalisation in construction.

2 Understanding the construction industry in developing countries

Introduction

The construction industry plays a pivotal role in the development of nations, and its contribution to socio-economic progress is enormous. The characteristics of the industry in developed countries vary considerably from that of developing countries. The consequence of this variation is that innovations being adopted in developed countries to develop construction services might not affect the industry in developing countries. The development of the construction industry in developed countries is faster than that of their counterparts in developing countries. Therefore, it is necessary first to explore the construction industry's characteristics in developing countries and understand the challenges facing the industry's development, before proposing innovative measures such as construction digitalisation to help the industry evolve. Therefore, this chapter explores the nature of the construction industry and its challenges in developing countries. Due to the complex and multifaceted nature of the construction industry, there is no unison in the body of literature regarding the construction industry's challenges (Elkhalifa, 2016). Exploring the industry's challenges in different countries as narrated in past studies gives a clear insight regarding some of the most critical issues that need to be addressed.

Furthermore, the chapter explores the construction industry's performance in relation to some predefined project success criteria. The concept of project success has been widely discussed in past studies, and it is believed to encompass so many factors. Determining whether a project has performed well, therefore, depends on the yardstick set for measuring success from the beginning of the project. Some of the prominent success measures are explored in this chapter, and the performance of the industry vis-à-vis these success measures are noted.

The construction industry

From time past, man has been raising structures in the form of shelters – an obvious need after food. As time went by, man's needs transcended from the need for just shelter to the need for more comfortable shelter with necessary facilities, coupled with the need for infrastructures within his immediate environment. The activity involved in creating these shelters and their associated facilities is known as "construction" (Wells, 1985). Construction has been described as every activity that leads to the creation and rectification of structures and facilities that are immovable. The provision of these immovable structures is made possible through the construction industry's expertise (Nam & Tatum, 1988). Therefore, the construction industry is vital to transforming the needs and aspirations of people into reality. The industry does this through the physical implementation of diverse construction projects (Ibrahim *et al.*, 2010).

The United States Census Bureau (2002) described the construction industry as an industry that primarily engages in constructing buildings and other structures. These organisations carry out heavy construction, reconstruction, alterations, and other construction-related activities. Chris (1998) and Gould and Joyce (2008) described the end product of the industry to be in two forms. The first is building works such as houses, hospitals, offices and business complexes, factories, etc. In contrast, the second is civil engineering-related works such as roads, bridges, infrastructure for water supply, power generation, etc. From the aspect of participants and operations needed to deliver these products, numerous parties, processes, phases, and stages are involved. There is a large amount of input from both the public and private sectors, with the primary aim of bringing the project to a successful conclusion. This makes construction a complex production of a unique product through the cooperation of multi-skilled ad hoc teams (Bertelsen & Koskela, 2004; Ibrahim *et al.*, 2010). Therefore, Navon (2005) concludes that the construction industry is complex because it comprises large numbers of parties and processes in the delivery of its products.

This complex nature and temporary coming together of construction participants are not without their disadvantages to the construction industry. Khalfan and Anumba (2000) opined that due to the fragmented nature in the realisation of projects within the construction industry, the industry all around the world has come under severe criticisms for its inefficiency and lack of productivity. Also, the traditional ways of delivering construction products have faced unprecedented challenges. There is a growing competition that forces construction organisations to rethink and re-evaluate their construction methods to improve productivity, quality, and efficiency (Karna & Jonnonen, 2005).

Despite these criticisms and complexity, the industry still contributes significantly to societal goals development and achievement. It has been noted that a nation's economic growth can be measured through the number of physical structures that are developed in the country (Ayodele & Alabi, 2011; Kazaz & Ulubeyli, 2009). Furthermore, urban and rural development goals can be achieved through the services delivered by the construction industry (Leibing, 2001). No country can attain its desired growth without considerable development and infrastructure built to spur the economy. This has made the construction industry necessary in every nation, not only in improving the economy but also in creating wealth and quality of life for the populace (Ibrahim *et al.*, 2010). Therefore, Ayodele and Alabi (2011) concluded that a healthy economy usually experiences an increase in activities of its construction industry.

In affirmation, the World Economic Forum (2016) submitted that the construction industry is crucial to economic development as all other industries (whether small scale or large scale) and sectors depend on it for the environment in which they operate. All over the world, the construction industry has come to be realised as a critical industry to national development (Construction Industry Development Board, CIDB, 2017). The importance of construction to economic growth and development cannot be overstressed as construction makes a significant contribution to gross domestic product (GDP), capital formation, and employment (Saka &Lowe, 2010). The industry accounts for up to 6% of the global GDP, adding a $3.6 trillion value, with nearly $10 trillion in the total annual incomes (World Economic Forum, 2016). In some developing countries, the industry's contribution to GDP can be up to 8% as in the case of India (Dixit *et al.*, 2017). In China, a contribution of 6.8% was recorded as of 2018, and it is expected that this figure will remain constant until 2021 (Statista Research Department, 2019). In Malaysia, the Department of Statistics noted a 4.2% contribution of the industry to the country's GDP in the third quarter of 2020 (Department of Statistics Malaysia, 2020). In Africa, a significant contribution is also witnessed in Kenya where the construction industry contributed 4.8% to the country's GDP in 2015, and this is expected to grow to 7% in subsequent years (Wetangula & Mazurewicz, 2017). In Nigeria and South Africa, the construction industry contributes 3.21% and 3.0% respectively to the countries' total GDP (Zingoni, 2020; National Bureau of Statistics, 2020).

The construction industry plays an important and dynamic role in any nation's sustainable economic growth as more than 50% of the gross fixed capital budget usually takes the form of construction output. Furthermore, employment in the sector is estimated at over 111 million workers all over the world (Durdyev & Ismail, 2012). For most developing countries, the construction industry is littered with small and medium enterprises which contribute significantly to job creation (Aigbavboa *et al.*, 2018). For example, the construction industry in South Africa contributes 11% to the country's total employment (CIDB, 2018). In Ghana and Nigeria, the industry makes up to 3% contribution to total employment (Ahadzie, 2019; Oladinrin *et al.*, 2012). A similar observation of 3.1% contribution was made in Zambia (Cheelo & Liebenthal, 2018). These jobs can be designated into management and specialised/skilled team, semi-skilled, and under-skilled workforce. The skilled workforce is usually employees with higher educational qualifications, who are taught to plan and run the construction process, and is predominantly part of the management and technical team. The semi-skilled workers are artisans, who include plumbers, bricklayers, welders, and plant operators. Lastly, unskilled workers are general labourers with less or no experience of how the construction industry works (Vitharana *et al.*, 2015).

Challenges facing the construction industry

The industry's role in ensuring the development of countries has been emphasised in past studies. However, despite this obvious importance of the industry, the construction industry in developing countries has not made any significant progress in recent times (Ofori, 2018). Highlighting the problems facing this industry in these developing countries becomes pertinent to the construction industry's development. Aptly, studies have shown that categorising the factors affecting the industry and hindering its development is difficult due to the industry's complex and multidimensional nature (Elkhalifa, 2016; Ofori, 2000). Furthermore, because the industry is a crucial driver of most countries' development, it is affected by situations within these countries; hence, the challenges might differ from one country to another. However, a look at past submissions shows some similarities in these challenges. Using Ghana as a case study, Ofori (1980) submitted eight factors which are still relevant to most construction industries in developing countries around the world today. These challenges include:

(1) Economic growth and stability
(2) Government recognition
(3) Planning and resources
(4) Codes and procedures
(5) Use of local materials
(6) Education and training
(7) Appropriate technologies
(8) Incentives for local contractors

Al-Omari (1992), in a case study conducted in the United Arab Emirates, discovered six major groups of problems facing the construction industry. These are;

(1) External environment
(2) Indigenous environment
(3) Development planning and policies
(4) Planning prerequisites and measurement tools
(5) Implementation strategies
(6) Working environment

Fox (2003) further modified these challenges and defined them into two groups. These are:

(1) Traditional factors

   • *Basic resources and institutional infrastructure*
   • *Financial and human resources*
   • *Techniques and technologies supporting high production*

(2) Cultural factors

   • *Long-term vision and policy for industry*
   • *Thinking the best and behaving the best*
   • *Learning culture*

Building on the aforementioned, Elkhalifa (2012) submitted four significant groups in assessing the construction industry in Sudan. These groups are:

(1) Socio-economic and political environment
(2) Construction industry environment

- *Construction industry structure and capacity*
- *Working environment and behavioural factors*

(3) Resources

- *Financial resources*
- *Physical resources*
- *Human resources*

(4) Supporting systems

- *Regulatory system and institutional framework*
- *Research and development*

Zawdie and Langford (2002) explored the Sub-Saharan Africa and submitted six major challenges facing the construction industry in delivering adequate infrastructure. These challenges are:

(1) Poor project financing
(2) Poor investment decisions
(3) Government participation
(4) Absence of private and foreign direct investment
(5) Weak engineering and management capabilities
(6) Lack of resources (manpower, materials, machines, foreign exchange)

According to Windapo and Cattell (2013), while the significant challenge facing the construction industry in South Africa is the rising cost of acquiring materials needed for building projects, several challenges also contribute to the lack of growth and development experienced by the industry. These challenges are:

- Increased costs of building materials
- Access to affordable mortgage/credit
- High interest rates
- The high rate of enterprise failure/delivery capacity and performance
- Mismatches between available skills and required skills
- Availability of infrastructure
- External influences such as government legislation
- Availability of suitable land
- Public-sector capacity
- Poverty
- Critical global issues/globalisation
- Procurement practices/capacity for sustainable empowerment
- Technology

In Nigeria, Ogunmankinde *et al.* (2019) submitted that the industry's challenges cut across major stakeholders, construction methods, and regulatory framework. The study further grouped these challenges into six distinct groups which are:

(1) Nature of the industry
(2) Clients
(3) Professionals
(4) Government
(5) Construction organisations
(6) Projects

In Ethiopia, Mengistu and Mahesh (2020) grouped these challenges deterring the growth of the country's construction industry into four. These are:

(1) The role of government
(2) Resource related
(3) Nature of the industry
(4) Industry's vision of its own development

Based on the aforementioned challenges, it is evident that the construction industry is facing critical issues in most of the developing countries around the world. These key challenges are discussed in the following.

Poor image of the industry

The construction industry faces significant challenges due to its fragmented nature which covers broad segments and diverse stakeholders and its poor image (Mengistu & Mahesh, 2020). The industry in developing countries is viewed as one that offers blue-collar jobs, which in most cases are believed to be low-status jobs (Haupt & Harinarain, 2016). These jobs are less appealing to the youths due to the need for manual labour and its associated low wages. As a result, young people prefer to seek college degrees to give them the opportunity for "high status" jobs (Tucker *et al.*, 1999). This poor image of the industry has been attributed to the nature of work done, which demands significant physical input and places the workers in some level of danger (International Labour Organisation, ILO, 2001). Also, the lack of a well-defined career path coupled with the industry's poor health and safety nature has been described as a reason for the poor image of the construction industry (Haupt & Harinarain, 2016; Makhene &Twala, 2009). This problem was ranked among the top ten challenges relating to the industry's nature in Ethiopia (Mengistu & Mahesh, 2020). In South Africa, the poor image of the industry tends to discourage students from picking up careers in the industry. The resultant effect of this lack of new entrants is the skills shortage that has bedevilled the construction industry in recent times (Haupt & Harinarain, 2016; Windapo & Cattell, 2010). Rameezdeen (2007) concluded that dedication and conscious effort are needed from all stakeholders in the construction industry if the image of the industry is to improve.

Health and safety

Though the construction industry is said to boost the economy, it also has several costly site accidents and injuries. While several efforts have been directed towards improving the health and safety (H&S) nature of construction industries worldwide, there is still an alarming rate of site accidents, injuries, and fatalities (Haupt & Harinarain, 2016). The situation is worse in developing and underdeveloped countries where the construction industry is considered one of the most dangerous industries (Aghimien, Oke *et al.*, 2018; Laryea & Mensah, 2010). In India, Nagarajan (2018) observed poor safety of workers as a problem faced by the construction industry. A similar observation was made in Eswatini (Aghimien, Aigbavboa, Thwala, & Thwala, 2019a) and South Africa (Haupt & Harinarain, 2016). This poor health and safety of the industry often leads to significant loss of production time and compromise of construction quality, vis-à-vis the associated pain and suffering of the victim's family, the reduction in staff morale as well as the high medical cost incurred (Aghimien, Oke *et al.*, 2018; Kadiri *et al.*, 2014). This situation portrays a poor image for the industry which invariably leads to a shortage of workers as potential workers get scared of the industry.

Lack of synergy

Aside from issues of poor image and H&S, the construction industry's nature has also been characterised by lack of synergy between stakeholders and some disconnect between professional bodies. These have led to significant failure in construction projects delivered in the industry (Aibinu & Odeyinka, 2006). Mengistu and Mahesh (2020) also noted this "disconnect" in academics and industry collaboration in the Ethiopian Construction Industry. A similar observation was made by Elkhalifa (2016) in Sudan as poor synergy between practice and research emanated as the second most significant factor affecting the construction industry. The resultant effect of this problem is the continuous failure of projects in the industry that could have been solved with a rigorous and practical research and development that the academic sector would have been able to deliver.

Unethical practices

Unethical practices are questionable behaviours that go against the professional norms of the different construction participants and the industry as a whole (Wold *et al.*, 2019). Issues surrounding construction participants' unethical practices have been pinpointed as the bane of the construction industry (Ho, 2011; Shah & Alotaibi, 2017). Often, cartels are formed by construction organisations to monopolise a sector or geographical region in most countries. With significant power garnered from this monopoly, tender prices are increased drastically (Haupt & Harinarain, 2016). The resultant effect of this corrupt activity is increased cost of delivery of construction projects. In South Africa, CIDB (2011) submitted that among several factors, corruption plays a significant role in the poor quality of construction projects being delivered in the country. In Ghana, Ameyaw *et al.* (2017) observed high levels of corruption and unethical behaviour among public officers, contractors, and consultants in the construction industry. A similar observation was made in Bahrain, India, Malaysia, and Pakistan (Transparency International, 2019). The situation is no different in Nigeria, where corruption tops the list of unethical practices in the country's construction industry. This is unsurprising as corruption is a general menace in the country which has found its way into the construction industry (Ogunmankinde *et al.*, 2019; Transparency International, 2019). Olatunji *et al.* (2016a) also submitted that there are issues of insincerity to clients and collusion among construction professionals while delivering construction projects.

Unethical practices are harmful not just to the construction industry but also to the economy and human resources of countries worldwide (Aigbavboa *et al.*, 2016; Olatunji *et al.*, 2016a; Oyewobi *et al.*, 2011). The occurrence of unethical practices gives the industry a bad reputation (Maseko, 2017). Such practices distort the entire construction process, hinder free play of market forces, and discourage investors from investing within the industry (Olatunji *et al.*, 2016a). Unethical practices such as corruption affect construction project performance adversely as it leads to unnecessary increase in construction cost and reduced quality of projects being delivered. The situation also slows the pace of economic development and creates social inequality, particularly in developing countries (Ameyaw *et al.*, 2017). As a result, a call for ethical practices within the industry has become a popular clamour (Vee & Skitmore, 2003).

Resistance to change and innovation

In India, Nagarajan (2018) argued that one of the significant challenges of the construction industry is that automation is replacing workers. While the study argued that the industry is still filled with small and micro enterprises who rely heavily on manual labour to deliver their projects, the submission points to the resistance to change in the construction industry. Similarly, the construction industry has been rigid with its "brick and mortar" principles. Accommodating new delivery methods is challenging, particularly when the current approach yields the expected result. According to De Soto *et al.* (2018), the construction industry is perceived as a "traditional" sector with an ageing workforce that is resistant to change; this puts the risk on return on investment into technologies and innovative processes since employees refuse to use them and gravitate to outdated processes that they feel comfortable with. This is done without realising that the adoption of technological tools will bode well for the industry.

Skill shortage

It has been established that the construction industry employs a significant number of people in the delivery of its services. These people are either employed temporary or permanently, and their contribution is pertinent to the successful realisation of the industry's end products. The shortage of needed skills in the industry has been a reoccurring issue in past studies (Elkhalifa, 2012; Windapo & Cattell, 2010; Zawdie & Langford, 2002). For example, in South Africa, Nel and Werner (2017) described the situation as a paradox. This description came from the country's high rate of unemployment, coupled with its scarcity of skilled workers in diverse industries and sectors. According to the Human Science Research Council (2009) this shortage of skills has significantly affected the country's rate of growth. In other countries such as India, Mozambique, and Nigeria, studies have pinpointed skill shortage as a crucial issue facing their construction industries (Ayedun & Oluwatobi, 2011; Nagarajan, 2018; Nhabinde *et al.*, 2012; Ogunman-kinde *et al.*, 2019; Sawhney *et al.*, 2014). To salvage this situation, investment in continuous training and development of the available skilled workforce is essential (Bustani, 2000).

Financial resources

The availability of financial resources will determine the financial capability and competitiveness of any organisation. However, the financial constraint of construction organisations in developing countries is a reoccurring one in past studies (Elkhalifa, 2016; Fox, 2003). Issues relating to financial incapability and high investment cost of production were identified as critical issues facing the construction industry in most African countries (Elkhalifa, 2016; Mengistu & Mahesh, 2020; Zawdie & Langford, 2002). Limited access to credit is a significant challenge facing construction organisations in Mozambique, resulting from the banking sector market concentration and low levels of financial savings (Nhabinde *et al.*, 2012). In Nigeria, lack of finance affects the construction industry adversely, as delays in projects, sub-standard delivery of projects, and project abandonment have been recorded over time (Chinedu & Fidelis, 2011). Due to the poor financial capability of construction organisations in developing countries (Idoro, 2009), foreign firms are now favoured with construction projects because they believe that they have the financial and technical capabilities required to deliver more satisfactory jobs (Famakin & Ojo, 2013). Furthermore, the continuous increase and instability in the price of building materials within these developing countries make it more difficult for clients to achieve their desired projects within their available budget (Elkhalifa, 2016; Windapo & Cattell, 2013). The result of this problem is the shortage of jobs and revenue for small and medium organisations who rely heavily on start-up capital from clients to execute construction projects (Odediran *et al.*, 2012).

Technology

Over time several factors have been held accountable for the poor performance of construction projects in developing countries. Among these factors is the slow pace of technological advancement. While technology is advancing rapidly, most of which offers significant benefits to the construction industry, construction organisations in developing countries have continuously failed to adopt these technologies to deliver their services (Oke *et al.*, 2018). Chilipunde (2010) observed that the slow adoption of technology and limited information technology (IT) skills affect contractors' productivity in Malawi. Fadhi and Tan (2001) also mentioned that the slow embrace of technological advancement is part of the factors responsible for the poor performance of construction works in construction industries worldwide. CIDB (2007) observed technology adoption issues as a crucial factor affecting the South African construction industry. Poor technology application was also rated among the top ten most significant resource-related challenges facing Ethiopia's construction industry (Mengistu & Mahesh, 2020). In Nigeria, Mbamali and Okotie (2012) observed that the increase in the application of technological approach among construction companies and professionals would go a long way in increasing construction projects' performance within the country.

Extant literature exists on the need for constant improvement in the conventional project delivery method in the construction industry. It has been revealed that building design was done traditionally with the use of simple tools until the advancement in building material science in the mid-nineteenth century when engineers began to use computers to produce 2D computer-aided designs. These designs are communicated via paperwork, and this has brought about several problems among construction participants. The result is the poor delivery of construction projects being experienced within the industry. Adopting innovative technologies that can aid in significant improvement in the delivery of construction projects within these developing countries is necessary. Digitalising construction process through the use of digital technologies can solve issues surrounding poor construction performance and ensure the successful delivery of construction services (Oke *et al.*, 2018; Yan & Damian, 2008).

Government support

The government is the biggest client, promoter, and regulator of the construction industry (Ogbu, 2017; World Economic Forum, 2016). Regrettably, the decline in government expenditure and the failure to pay contractors promptly have contributed to the decline in the construction industry in developing countries (African News Agency, 2019). Milford (2009) mentioned that since public funds in most developing countries are significantly strained, construction industries in these countries face reduced demands. Also, the investors' confidence in the construction industry depends mostly on the government's ability to deliver infrastructure on the planned delivery programme through the industry. However, the successful delivery of these infrastructures is usually limited, thus affecting investors and businesses (CIDB, 2007).

Economic conditions and poverty

The issue of poverty is a reoccurring menace in most developing countries. It has been observed that significant numbers of people in developing countries worldwide live in extreme poverty and malnourishment (Mbande, 2010; Sanchez *et al.*, 2005). Many of these people exist in Asia and sub-Saharan Africa. In most of these countries, particularly in Africa, the poverty level is on a steady rise as the British Broadcasting Corporation (BBC) in 2012 reported that poverty in Nigeria has risen with almost 100 million people surviving on less than a dollar per day (BBC, 2012). It was observed that despite the increase in economic growth, the citizens have remained significantly poor (Okhiria & Obadeyi, 2015). In Southern Africa, almost 50% of the people who are largely residing in rural areas live in poverty with less than a dollar per day (African Development Bank, 2004). Stats SA (2017) reported that despite the poverty decline experienced between 2006 and 2011, there was an increase in poverty levels in 2015. George and Ukpong (2013) noted that some of the major causes of poverty include lack of employment, high dependent population, and overpopulation. With overpopulation comes increased pressure on the meagre available infrastructure needed for survival. This leads to all other negative happenings within society. Muhammad (2012) described poverty as the "*octopus dragging along multiple criminal and nefarious activities among young and old, male and female folks*".

Alleviating the poverty issue is crucial for every country's development and by extension, the development of its various industries of which the construction industry is included. Windapo and Cattell (2013) observed that viewing poverty from afar might tempt one to ask what it has to do with the construction industry. However, understanding that poverty can weaken the world economy and also create global unrest, alleviating it became one of the major Millennium Development Goals (van Wyk, 2004). Mbande (2010) further noted that since the proclamation of these goals to tackle poverty, many donor nations are linking their funding of infrastructure to the achievement of socio-economic goals. Even the newly proclaimed Sustainable Development Goals is centred around poverty eradication and recognises the need for infrastructure development as a significant goal. Windapo and Cattell (2013) concluded that having access to infrastructure development funds can prove to be a useful tool in the development of the construction industry.

Performance of the construction industry

Changes in the demands of clients, increasingly complex nature of construction designs, and advancement in emerging technologies have made the need for the development of construction industry services a must in countries worldwide. However, this development has been slow in developing countries (Ofori, 2018) due to some of the challenges previously mentioned. As a result, the construction industry in developing countries has been characterised with poor project performance. This performance has been measured using diverse success criteria. Studies have shown that while the trio of cost, time, and quality has been favoured over the years, project success measures transcend these three criteria (Atkinson, 1999; Atkinson *et al.*, 1997; Oke & Aigbavboa, 2017). Atkinson (1999) argued that cost, time, and quality are two best guesses and a phenomenon. The cost and time dimensions of project success are described as quantitative variables which do not capture other subjective and non-quantitative criteria (Chan, 2001; Takim & Akintoye, 2002).

Furthermore, the quality dimension is believed to be a subjective view of projects whose rating depends on the rater's opinion (Giusca et al., 2009). Therefore, following Atkinson (1999) suggesting the need for new success criteria, different views of the success of construction projects have continued to emerge. These measures have been classified into other groups such as objective and subjective measures, micro and macro measures, delivery and post-delivery measures, internal and external measures, short- and long-term measures, qualitative and quantitative measures, etc. (Atkinson, 1999; Lim & Mohamed, 1999; Oke & Aigbavboa, 2017; Shenhar et al., 1997; Silva et al., 2016). Some of these success criteria include health and safety, client and stakeholder's satisfaction, managerial, financial, technical, and organisational performance, sustainability, employees' satisfaction, cash-flow management, and profitability.

Cost

Cost is a crucial element to determine the success of construction projects. This is because construction clients are most concerned with the cost they will incur in the realisation of their dream structure. Construction cost is a vital measure of project success throughout the project's lifecycle, and it is of great concern to construction participants (Hussin et al., 2013). The construction cost is mostly estimated at the preconstruction phase of the project, and it is centred around the client's budget. In practice, several estimates are made to give the client a view of what will be spent on the project. Mostly, a preliminary estimate is done to establish a benchmark for the project at the earlier stage of the project, while a detailed cost estimate will be developed once all designs are ready, to show clearly the final cost of the project (Ashworth, 2004). The developed detailed estimate gives the target price, which is agreed upon before the planning and actual construction begins (Oke & Aigbavboa, 2017). Clients reasonably expect that the project is completed within the agreed target price. When this happens, it is believed that such a project has performed in terms of cost. Therefore, a construction project's cost performance is measured by comparing completion cost against budget or initial cost. This includes considering the various adjustments such as variations, fluctuations, design modifications, and additional works that might have been made during construction. Bubshait and Almohawis (1994) describe cost performance as the degree to which the general conditions promote completing a project within the estimated budget. By comparing the completion cost with the estimated budget, the cost variance can be determined to identify the extent to which the completion cost deviates from the budgeted cost. According to Salter and Torbett (2003), cost variance is the most common technique used to measure project performance. It is not only confined to the tender sum (i.e., the overall price provided by the contractor for which the contractor is willing to do the job) but also the overall cost that a project incurs from inception to completion, which includes any cost arising from variation, modification during the construction period, and the cost arising from legal claims such as litigation and arbitration (Chan & Chan, 2004).

While cost has been emphasised as a crucial measure of project success, most construction projects' performance in this regard has been below expectation of clients. Ogunsemi (2015) submitted that it is rear to see a construction project delivered within its estimated budget. In most countries, nine out of ten projects overrun their budgeted cost (Flyvbjerg *et al.*, 2003), and this has been a serious problem for developing countries where these overruns sometimes exceed 100% of the anticipated cost of the project (Azis *et al.*, 2013; Memon, 2013). This cost overrun is evident in diverse projects such as civil engineering projects (Hussin *et al.*, 2013; Le-Hoai *et al.*, 2008), building structures (Endut *et al.*, 2009; Jackson, 2002; Zujo *et al.*, 2010), among others. In Croatia, 81% of 333 projects suffered from cost overrun (Zujo *et al.*, 2010), while in Bosnia and Herzegovina, about 41% of 177 projects suffered a cost overrun (Zujo & Car-Pusic, 2008). In Kuwait, 33% of housing projects recorded cost overrun (Koushki *et al.*, 2005). In Malaysia, more than half of the public and private projects delivered in the country experienced cost overrun (Endut *et al.*, 2009; Potty *et al.*, 2011). In India, Singh *et al.* (2009) recorded an average of 15.8% of cost overrun in the projects evaluated, while Lee (2008) submitted 50% cost overrun in rail projects executed in Korea. In Nigeria, Olatunji (2008) submitted that the country's rate of cost overruns is alarming. Similarly, cost overrun ranging from 4.78% to 34.7% has been observed in past studies emanating from the country (Aghimien & Awodele, 2017; Omoregie & Radford, 2006). Also, in Zambia, Kaliba *et al.* (2009) recorded 69% cost overrun for road projects.

No doubt, these excessive cost overruns incurred in the delivery of construction projects have negative implications for the stakeholders, the industry, and society (Mbachu & Nkando, 2004; Olatunji *et al.*, 2016b). From the client's perspective, high cost implies added costs above those initially agreed upon at the onset, resulting in less investment return. From the consultant's perspective, it means the inability to deliver value for money and the possibility of damage to reputation and loss of confidence among clients. The contractor's perspective implies a loss of profit through penalties for non-completion and negative word of mouth that could jeopardise chances of winning future jobs. From the end user's view, the added cost is passed on as higher rental/lease costs or sales prices (Ogunsemi, 2015). Based on the preceding, it is clear that the construction industry needs to develop innovative solutions that will help solve the problem of poor cost performance and improve the industry's reputation.

Perhaps, highlighting some of the factors responsible for these cost overruns can help give better direction as to the solutions needed. Since cost overrun is common among different countries, different aspects have emanated as the bane of this problem. For example, in India, Iyer and Jha (2005) submitted factors such as project manager's competence, top management support, project manager's coordinating and leadership skill, monitoring and feedback by the participants, decision-making, project participants' coordination, owners' competence, social, economic, and climatic condition. In Palestine, Enshassi *et al.* (2009) submitted crucial factors such as the escalation of material price, cash flow, material and equipment cost, and liquidity of an organisation. In Nigeria, Ameh *et al.* (2010) discovered lack of experience of contractors, cost of material, fluctuation in the prices of materials, frequent design changes, economic instability, high-interest rates on loans and mode of financing, bonds and payments as well as fraudulent practices as the major factors leading to cost overrun in construction projects.

In Ghana, Frimpong *et al.* (2003) submitted factors such as stage payment difficulties by clients, poor control of the contract, material procurement, poor technical performances, and escalation of material prices. Also, in South Africa, Mulalo *et al.* (2018) submitted that the major factors causing poor cost performance are poor planning, material price fluctuation, high-interest rates on loans, inflation, increase in the cost of hiring labour, poor coordination between construction participants, poor financial management on projects, as well as inadequate production of raw materials locally. In Malaysia, Memon *et al.* (2012) categorised these factors into seven. These are contractor's site management, design and documentation, financial management, information and communication, human resource, non-human resource, project management, and contract administration-related factor.

Looking at past studies that have assessed the factors responsible for the poor performance of construction projects in terms of cost, it is clear that a myriad of factors exists. Prominent among these factors are:

- The fluctuation of prices of materials
- Poor site management and supervision
- Incompetence and lack of experience of construction participants
- Inadequate planning and scheduling
- Unforeseen circumstances
- Inaccurate cost estimates at the initial stage of the project
- Mistakes during construction and frequent design changes
- Inadequate monitoring and control
- Incomplete design at the time of tender
- Cash flow and financial difficulties faced by clients and contractors
- Poor financial control on site
- Delay in progress payment by the owner
- Contractual claims, such as the extension of time with cost claims
- Lack of communication and coordination between parties
- Poor labour productivity
- Shortage of workers leading to the high cost of labour
- Shortages of materials and late delivery of materials and equipment
- Poor project management
- Change in the scope of the project
- Delays in decision-making
- Fraud and corruption
- Project complexity

- Poor technology adoption
- Inconsistency in government policies and industry regulations

Time

Time overrun has also become a crucial problem for most construction projects where clients and stakeholders evaluate success based on timely completion (Lim & Mohamed, 1999). For some clients, delivering a project within the allotted time equals timely revenue from such projects. Thus, failure to meet this schedule amounts to a loss of revenue for such clients and subsequently, project failure (Salter & Torbett, 2003). Unfortunately, achieving project success in terms of the agreed schedule has been a severe problem for the construction industry (Ogunsemi & Jagboro, 2006). Most projects delivered within the industry has been characterised by poor performance with time and its attributed cost implication on projects (Chan & Chan, 2004).

Time or schedule in construction refers to the duration of completing a project. This is the absolute time calculated in days/weeks starting from the day the site work commenced to the project's practical completion (Chan & Chan 2004). Silva *et al.* (2016) described construction time as the project's agreed period to be completed. Oke and Aigbavboa (2017) described this criterion as a crucial traditional factor used in determining project success, and it is also a critical factor in selecting contractors in most projects. This is because contractors are made to submit project duration during tendering, and this duration is evaluated in line with the client's expectation and requirement. Like cost, the construction project's time performance is evaluated by assessing the variation between the estimated time from the initial stage of the project and its completion time. The importance of construction projects being delivered to agreed time cannot be overemphasised (Lim & Mohammed, 1999). But just like cost, it is rare to find construction projects that are free of time overrun. In Jordan, Al-Momani (2000) discovered 82% of 130 public projects experienced time overrun. In Ghana, Frimpong *et al.* (2003) recorded 70% of 47 projects delivered behind schedule. The story is not different in Nigeria, where 1,517 projects were found to have experienced schedule delay in their delivery (Amu & Adesanya, 2011). Aghimien and Awodele (2017) also found 130% time overrun, which is similar to the 188% earlier mentioned by Omoregie and Radford (2006).

Past studies have mostly categorised possible factors causing a delay in project completion into four major categories: the consultant's responsibility, contractor's responsibilities, client responsibilities, and external factors (Chan *et al.*, 2004; Ogunlana *et al.*, 1996). Abu-Shaban (2008) suggested that the factors that could influence the time performance of a construction project in Palestine include site preparation time, planned time for project construction, percentage of orders delivered late, the time needed to implement variation orders, the time required to rectify defects, the average delay in claim approval, average delay in payment from the owner to the contractor, availability of resources as planned through project duration, average delay because of closures and materials shortage.

Kikwasi (2012) opined that since construction projects are mostly time-bound, proper time management is necessary for eliminating all avenues of delays. In understanding the causes of poor time performance of construction projects, several factors have been unearthed in past studies. For example, Noulmanee *et al.* (1999) investigated causes of poor time performance in highway construction in Thailand and concluded that this could be caused by all parties involved in projects; however, the leading causes are inadequacy of subcontractors, lack of sufficient resources, incomplete and unclear drawings, and deficiencies between consultants and contractors. In Jordan, Al-Momani (2000) observed designers' issues, changes from clients, weather and site conditions, late deliveries, economic conditions, and increase in quantity.

Al-Kharashi and Skitmore (2008) pointed out that the leading cause of poor time performance in Saudi Arabia construction sector is the lack of qualified and experienced personnel. In Malaysia, Sambasivan and Soon (2007) identified ten most important causes of poor time performance of construction projects as contractor's improper planning, contractor's poor site management, inadequate contractor experience, inadequate client's finance and payments for completed work, problems with subcontractors, shortage in material, labour supply, equipment availability and failure, lack of communication between parties, and mistakes during the construction stage.

Chan and Kumaraswamy (1997) identified five principal factors inhibiting time performance: poor risk management and supervision, unforeseen site conditions, slow decision-making, client-initiated variations, and work variations. Haseeb *et al.* (2011) pointed out that the most common factors of poor timely delivery of construction projects in Pakistan are natural disasters like floods and earthquakes. The study also acknowledged others factors such as financial and payment problems, improper planning, poor site management, insufficient experience, and shortage of materials and equipment.

In Nigeria, Odeyinka and Yusif (1997) assessed the causes of delays which invariably lead to poor time performance in building projects. They classified these causes as project participants and extraneous factors. Client-related factors included variation in orders, slow decision-making, and cash flow problems under the project participants. Contractor-related factors identified were financial difficulties, material management problems, planning and scheduling problems, inadequate site inspection, equipment management problems and shortage of workforce. The extraneous factors identified were inclement weather, acts of nature, labour disputes, and strikes. In a similar vein, Ubani *et al.* (2013) categorised factors that inhibit time performance of construction projects into seven major groups: client, contractor, consultant, materials, labour and equipment, contractual relationships, and external factors. On a general view, the factors affecting the time performance of construction projects in developing countries include:

- Poor finance and delay in payment of completed work
- Client interference
- Slow decision-making by clients

- Unrealistic imposed contract duration
- Subcontractors incompetence
- Poor site management
- Poor and inaccurate construction methods
- Improper planning
- Mistakes during construction
- Inadequate contractor experience
- Contract management
- Preparation and approval of drawings
- Quality assurance/control
- Waiting time for approval of test and inspections
- Quality of material
- Shortage in material
- Labour supply
- Labour productivity
- Equipment availability and failure
- Major disputes and negotiations
- Inappropriate organisational structure linking parties
- Lack of communication between the parties
- Weather condition
- Regulatory changes and building code
- Unforeseen ground conditions
- Political conditions
- Economic conditions

Quality

Quality has been described as one of the traditional measures in determining the performance of construction projects. Khosravi and Afshari (2011) described quality as the projects' ability to meet the agreed technical specifications. Other studies have distinguished between quality, technical performance, functionality, and the quality of the process and that of the product (Al-Tmeemy *et al.*, 2011; Chovichien & Nguyen, 2013; Heravi & Ilbeigi, 2012). Despite these differences, evidence abounds that quality, technical performance, functionality, and specification attainment are interconnected. This interconnectivity can be traced to the fact that the quality of construction products and the processes used in achieving the product can be traced to the functionality and technical performance of the product (Chan *et al.*, 2002; Silva *et al.*, 2016). Based on this knowledge, Oke and Aigbavboa (2017) described the quality of construction projects as a subjective criterion which is a "*combination of attributes that are expected of the services required by participants as well as elements or components of construction projects*". It was further stated that quality is related to the project's specification and the conditions of the contract. Thus, a project is considered to have performed in terms of quality when it is delivered in line with the outlined specifications and the conditions of the contract (*ibid.*).

Studies have shown that poor quality in construction projects has become a common sight for most developing and underdeveloped countries (Ali & Wen, 2011; Kazaz & Birgonul, 2005; Oyedele et al., 2015). Several factors have been held accountable for this shortcoming in the quality of construction projects (Arditi & Gunaydin, 1997; Dozzi et al., 1996; Kam & Tang, 1998; Love & Irani, 2003; Obiegbu, 2003; Oyedele et al., 2015; Oke & Aigbavboa, 2017; Serpell & Alarco'n, 1998). Some of these factors are:

- Errors in drawings and specifications
- Lack of adherence to quality standards
- Lack of qualified expertise to handle specialised aspects of designs
- Poor project planning
- Poor project monitoring and control from inception to completion
- Conflict among project participants
- Use of wrong or odd conditions of contracts
- Lack of proper risk assessment right from the early stage of the project
- Continuous alteration of specification instructions and conditions of contracts
- Poor storage of materials and equipment
- Strictness on delivering a project to cost and within schedule
- Poor communication and coordination among construction participants
- Emphasis on the "lowest bidder wins" concept
- Construction participants' greed and corruption

Stakeholders' satisfaction

In construction, several stakeholders exist. This includes the commissioned body (the client, owner, or sponsor), the construction participants involved in the realisation of the project, as well as the end-users of the project (Oke & Aigbavboa, 2017). When a client decides to select a particular contractor to provide construction services, expectations are formulated as to the result based on defined criteria. For a client to be satisfied, the contractors' performance must surpass or meet up with the set expectations (Folorunso & Awodele, 2015; Mbachu & Nkado, 2006). While the client is the principal entity in the conceptualisation of projects and perhaps the most important stakeholder, the satisfaction of construction participants and end-users is equally essential in evaluating construction projects' performance. This is evident in past studies that have emphasised the need for participant satisfaction, user satisfaction, client satisfaction, and customer satisfaction as various perspectives of viewing stakeholder's satisfaction in projects (Heravi & Ilbeigi, 2012; Khosravi & Afshari, 2011; Takim & Adnan, 2008). Therefore, contractors need to assess stakeholders' needs carefully. These needs can be broadly categorised into the design, management, and services, as noted by Bennett's (1985). Building on this Folorunso and Awodele (2015) submitted that these needs could be assessed from adherence to schedule, adherence to budget, quality of construction and workmanship, safety measures and standards, resources management, site personnel, quality of service, and attitude. Karna *et al.*

(2004) mentioned that satisfaction could be derived from five dimensions: quality assurance and handover, environment and safety at work, cooperation, personnel, site supervision, and subcontracting. Omonori and Lawal (2014) noted that satisfaction could be measured along the traditional line of quality, price, and schedule. A similar notion was held by Nzekwe-Excel (2007), who emphasised quality as a measure of satisfaction. However, this project success dimension (stakeholder satisfaction) has equally been seen as a challenge for quality improvement in the construction industry (*ibid.*). To improve this situation, critical factors such as careful management of project finance, using the right set of skilled workers, employing up-to-date technologies, ensuring quality customer relation, and effective time management should be considered as they tend to influence the satisfaction of stakeholders (Alshihre *et al.*, 2020).

Health and safety

Highlights of studies show high rates of accidents, injuries, and fatalities in the construction industry and construction site employees are at increased risk as they are the hub of construction activities (Bust *et al.*, 2008; Chiocha *et al.*, 2011; Okoye *et al.*, 2017; Othman *et al.*, 2017). Laryea and Mensah (2010) mentioned that the construction industries in underdeveloped and developing countries are performing below standard as poor material and nonconformance to safety leads to site accidents, collapsing of buildings, and costly repairs being made. Puplampu and Quartey (2012) further described the African continent as one with significant issues relating to H&S management. This is because H&S is not given adequate consideration in the quest for infrastructure development. Oversight of essential H&S management practices is common within the continent's construction industries (Orji *et al.*, 2016). Muiruri and Mulinge (2014) submitted that considering the alarming rate of construction site accidents, it is clear that safety acts either do not exist or are handled with "kids' glove" in terms of enforcement.

In most countries, the poor H&S status is not unconnected to negligence on the part of construction participants, fictitious or skewed accident records and wrong statistics, together with poor statutory regulations (Bust *et al.*, 2008). Kheni *et al.* (2008) identified lack of comprehensive resources, skilled employees, finance, and government support to Occupational Health and Safety (OHS) regulatory bodies as some of the problems of proper H&S practices in construction. Puplampu and Quartey (2012) similarly noted a lack of firm national OHS policy in Ghana, with poor H&S culture experienced within the construction industry. In Kuwait, Kartam *et al.* (2000) submitted that the lack of government effort to help regulatory bodies by injecting capital and giving H&S serious considerations affects the adoption of proper H&S practices. Another critical factor is the clients' attitudes and involvement in H&S practices. This is because most clients have no budget for insurance of H&S cases on their construction projects. According to Wadick (2010), construction clients are more concerned about the finished product than the H&S of workers delivering them. Mashwama and Musonda (2014) further stated that since improving H&S performance is seen as contractors' sole responsibility, clients do not see the need to encourage these contractors by supporting them financially. As noted by Loosemore *et al.* (1999) and Smallwood (2004), contractors who emphasise H&S are in most cases in danger of losing jobs to those who do not, due to the increased associated cost. This forces most SMEs struggling to stay afloat to discard some H&S practices in a bid to retain the goodwill of their clients and obtain more jobs (Aghimien, Oke *et al.*, 2018; Ozmec *et al.*, 2015). Therefore, construction clients could influence OHS throughout the different stages of construction project delivery if they understand its effectiveness in their projects' timely delivery (Lopes *et al.*, 2011).

Various studies show inadequate understanding and recognition of H&S as a challenge in most small and medium organisations in developing countries. The ILO (1998) records proved no information on records and statistics of accidents not because accidents do not happen on sites. This is evidence that there is a lack of substantial knowledge of the importance of keeping records. Ferreira and van Loggerenberg (2011) stated that construction workers themselves are responsible for their safety, although out of ignorance, they compromise site safety. Construction workers fail to report to their bosses about the problems they face concerning H&S in their construction environments. This makes their management assume that there are no safety issues and see no point in implementing or introducing mitigating measures. Attabra-Yartey (2012) also noted that most construction organisations have failed to make H&S a primary concern and this is sometimes as a result of some contractor's inadequate expertise, ignorance, lack of encouragement, and insufficient capital to invest in H&S. In affirmation, Sunindijo (2015) stated that most organisations' H&S performance is affected by a lack of financial stability. Lack of resources causes them to fail to budget for H&S management implementation. These financial constraints tend to make construction organisations compromise in hiring illegal migrants whom they treat as cheap labour (Rogerson, 1999). With these cheap labours, it is very easy to compromise H&S because even if they get harmed, there will be no legal costs and records.

Similarly, in a bid to reduce cost, most construction organisations rely on casual workers with less training to deliver projects. These casual workers have no employment protection, and their jobs are not secured because of their employment conditions, and earnings and obtainability are undetermined. Their training is deemed unnecessary because management prefers to train permanent workers (if they have any) (Fourie, 2008). It is a general norm that business is established for the sake of making profits. The success of construction business rests upon making more profit from its projects. According to Spellman (2016), many contractors remove the H&S allocations or budgets (including hiring the right workforce and training existing ones), whenever they intend to cut costs and maximise their profits.

Issues surrounding management commitment to H&S practices have also been noted. According to Chiocha *et al.* (2011), poor H&S practices can be due to senior management's inadequate dedication to H&S caused by lack of understanding of the inherent benefits that await both the SMEs and construction industry when these practices are appropriately adopted. Similarly, Muiruri and Mulinge (2014) noted that senior management is not committed to H&S in construction sites in developing countries like Kenya. This lack of commitment on the part of manage-ment leads to poor training of their employees in the aspect of H&S. Training and quality are connected to achieving H&S standards. Through proper H&S train-ing, organisations' reputation can improve, reduced costs and time wastage can be achieved, legal complications and disruption in the series of construction activities can be eliminated (Attabra-Yartey, 2012).

The term "survival of the fittest" that has characterised the construction mar-ket has compromised H&S as clients consider giving jobs to the lowest tenderer as an excellent system to ensure the reduced cost of construction. In this way, con-tractors tend to lower their costs and win more tenders (Torbica & Stroh, 2001). Afterwards, they handle too many jobs at a time, and they do not have adequate time for H&S proper planning, implementing, and controlling. This situation leads to them working long hours to meet up with strict deadlines. This affects the H&S of their workers, as observed by Aghimien, Oke et al. (2018). Similarly, risky behaviours on the part of construction workers are seen to be contributing to H&S issues within the construction industry (Oswald et al., 2013). This has a sig-nificant influence on H&S management practice. Unsafe behaviour such as work-ing with moving machinery, wearing dangling clothes, dangerous lifting, carrying and placing, the influence of alcohol and drugs were noted among construction site workers in Sri Lanka (Vitharana et al., 2015). A similar observation was made by Okoye et al. (2017) in Nigeria.

Sustainability

All around the world humans are living beyond their means with excess use of energy and significant wastage of natural resources (Hendricks, 2016). These human activities on the environment birth the emerging movement of sustainable development. The Meadows, Randers, and Behrens report in 1972 was one of the critical works highlighting this thinking. The Brundtland Report in 1987 pro-vided sustainable development as the solution to the problems that human activi-ties pose to the environment. The report gave the universally accepted definition of sustainable development as the "*development that meets the needs of the present without compromising future generations' ability to meet their own needs*" (Brundtland Report, 1987).

For sustainable development to be achieved, there must be sustainable con-struction as construction is a crucial driver of environmental development. However, as much as the construction industry contributes to every country's socio-economic development, the industry's activities also affect the environ-ment adversely (Aghimien, Aigbavboa, & Thwala, 2019). This adverse effect cuts across excessive exploitation of the environment for raw materials, land degra-dation, high energy consumption down to pollution from construction activities and waste generation (Ding & Langston, 2004; Griffith et al., 2005; Sjostrom & Bakens, 1999). As a result, there has been a clamour for sustainability in deliv-ering construction products among researchers and professionals in the industry

(Ametepey & Aigbavboa, 2014). However, viewing construction sustainability from the environment perspective alone has been considered a myopic view of the concept (Oke & Aigbavboa, 2017). The social and economic impact of construction projects needs to be considered to have a holistic view of construction projects' sustainability (Beheiry, 2006; Oke *et al.*, 2015). Therefore, sustainable construction has been proposed as an ideal way of making construction activities and project delivery more economically, socially, and environmentally sustainable (Abidin, 2010; Aghimien, Aigbavboa, & Thwala, 2019). The construction industry must be ready to jettison the old linear ways of delivering projects and embrace a more cyclic approach where 'life to life' process is a norm. This means that construction activities need to change and improvements that are compatible with the environmental, social, economic, and other limits both in the present and long-term future must be encouraged (Gray & Wiedemann, 1999).

There is no gainsaying that construction projects being delivered in developing countries are underperforming in terms of sustainability. Evidence abounds in this regard (Alabi, 2012; Al-Sanad, 2015; Aghimien, Adegbemo *et al.*, 2018; Baron & Donath, 2016). To address this issue, the focus has been placed on different aspects of sustainability related issues in past studies. Issues surrounding the perception, awareness, and ways of improving the sustainability delivery of projects, challenges facing this concept, use of renewable energy sources, green buildings, as well as management tools that can help promote sustainability have been explored (Abolore, 2012; Aghimien, Aigbavboa, & Thwala, 2019; Ahmed & Gidado, 2008; Aigbavboa *et al.*, 2017; Ametepey *et al.*, 2015; Djokoto *et al.*, 2014; Isa *et al.*, 2013; Oke *et al.*, 2015).

Several factors deterring the attainment of sustainability concept have emerged (Abidin *et al.* 2003; Aghimien, Adegbemo *et al.*, 2018; Aghimien, Aigbavboa, & Thwala, 2019b; Al-Sanad, 2015; Alabi, 2012; Akbiyikli *et al.*, 2009; Aigbavboa *et al.*, 2017; Ayarkwa *et al.*, 2017; Baron & Donath, 2016; Djokoto *et al.*, 2014; Kibert, 2013; Lowe & Zhou, 2003; Mousa, 2015; Opoku & Ahmed, 2014; Osaily, 2010; Powmya & Abidin, 2014), and they provide possible directions on the way forward in attaining sustainability in construction projects delivered in most developing countries. Some of these factors are related to construction, awareness and knowledge, finance, government, and technology (Aghimien, Adegbemo *et al.*, 2018). The factors include:

- Industry's resistance to change
- Client's preference
- Fear of increased cost of investment
- Inadequate knowledge and understanding of the concept of sustainability
- Fear of high investment cost
- Limited access to relevant information and historical data
- Inadequate government policies/support
- Lack of expert opinions on sustainable construction
- Inadequate exemplar projects
- Inadequate technology and technological process

- Lack of commitment of top management
- Unstable prices of construction materials
- Limited availability of sustainable materials in the local market
- Low demand level for sustainable products
- Inadequate sustainability measurement tools
- Inadequate building codes and regulations on sustainability
- The perception that sustainable construction materials are of low status
- Poor level of integration of life cycle cost
- Poor workmanship during construction
- Mode of funding of the project

Summary

The construction industry in developing countries has failed to develop despite its influence on the growth of these nations through substantial contribution to their GDP, provision of employment, and infrastructure delivery. The industry is faced with challenges which deter its development. Some of these challenges are evident in the industry's poor image, poor health and safety, lack of synergy between industry participants, unethical practices, resistance to change and innovation, lack of human, financial, and technological resources, unfavourable government policies, economic conditions, and poverty among others. These challenges further affect the ability of the construction industry to deliver successful projects. It is uncommon to see construction projects delivered within the estimated budget and schedule and to the predefined specification. Health and safety delivery of most projects is poor. In most cases these projects are not economically, socially, or environmentally sustainable, thereby leaving clients and other stakeholders dissatisfied at the end. Based on the foregoing, it is evident that construction digitalisation is necessary for construction industries in developing countries if the issue of poor performance of projects is to be solved.

References

Abidin, N.Z. (2010). Investigating the awareness and application of sustainable construction concept by Malaysian developers. *Habitat International*, 34(4): 421–426
Abidin, N.Z., Khalfan, M. and Kashyap, M. (2003). Moving towards more sustainable construction. *Proceedings of the Construction and Building Research Conference of the Royal Institution of Chartered Surveyors*, School of Engineering and the Built Environment University of Wolverhampton, Wolverhampton, 1–2 September
Abolore, A.A. (2012). Comparative study of environmental sustainability in building construction in Nigeria and Malaysia. *Journal of Emerging Trends in Economics and Management Science*, 3(6): 951–961
Abu-Shaban, S.S. (2008). *Factors Affecting the Performance of Construction Projects in the Gaza Strip*. An unpublished M.Sc thesis, The Islamic University of Gaza, Palestine
African Development Bank (2004). Bank Group Policy on Poverty Reduction. *African Development Group*. Available at: https://www.afdb.org/en/topics-and-sectors/topics/poverty-reduction#:~:text=Bank%20Group%20Policy%20on%20Poverty%20Reduction

%2C%20February%202004&text=The%20main%20objective%20of%20the,Bank%20 support%20for%20its%20RMCs.&text=The%20policy%20also%20ensures%20 consistency,the%20participation%20dimensions%20of%20PRSPs [accessed 20–8–2020]

African News Agency. (2019). SA govt must help stimulate construction industry – industry official. *Business Report*. Available at: www.iol.co.za/business-report/economy/sa-govt-must-help-stimulate-construction-industry-industry-official-36306857 [accessed 20–8–2020]

Aghimien, D.O., Adegbemo, T.F., Aghimien, I.E. and Awodele, A.O. (2018). Challenges of sustainable construction: A study of educational buildings in Nigeria. *International Journal of Built Environment and Sustainability*, 5(1): 33–46

Aghimien, D.O., Aigbavboa, C.O., Thwala, G. and Thwala, W.D. (2019). Impediments to health and safety management in SMEs in the Swaziland construction industry. *1st Association of Researchers in Construction Safety, Health, and Well-Being (ARCOSH) Conference*, Cape Town, South Africa, 3–4 June, pp. 115–127

Aghimien, D.O., Aigbavboa, C.O. and Thwala, W.D. (2019). Microscoping the challenges of sustainable construction in developing countries. *Journal of Engineering, Design and Technology*, 17(6): 1110–1128

Aghimien, D.O. and Awodele, O.A. (2017). Variability of cost and time delivery of educational buildings in Nigeria. *International Journal of Built Environment and Sustainability*, 4(3): 156–164

Aghimien, D.O., Oke, A.E., Aigbavboa, C.O. and Ontlametse, K. (2018). Factors contributing to disabling injuries and fatalities in the South African construction industry. *Joint CIB W099 and TG59 International Safety, Health, and People in Construction Conference*, Held in Salvador, in Brazil in 1–3 of August, pp. 337–345

Ahadzie, D.K. (2019). Ghana's construction industry is lively but needs regulation. *The Conversation*. Available at: https://theconversation.com/ghanas-construction-industry-is-lively-but-needs-regulation-124733#:~:text=In%202016%20and%202017%2C%20 construction,3%25%20of%20the%20labour%20force [accessed 20–8–2020]

Ahmed, A. and Gidado, K. (2008). Evaluating the potential of renewable energy technologies for buildings in Nigeria. In Dainty, A. (Ed.), *Proceedings 24th Annual Conference of Association of Researchers in Construction Management*, ARCOM, University of Glamorgan, Cardiff, Wales, 1–3 September

Aibinu, A.A. and Odeyinka, H.A. (2006). Construction delays and their causative factors in Nigeria. *Journal of Construction Engineering and Management*, 132(7): 667–677

Aigbavboa, C.O., Aghimien, D.O., Oke, A.E. and Mabasa, K. (2018). A preliminary study of critical factors impeding the growth of SMMES in the construction industry in Lusaka, Zambia. *Proceedings of the 3rd North American International Conference on Industrial Engineering and Operations Management*, Washington, DC, September 27–29, pp. 100–107

Aigbavboa, C.O., Ohiomah, I. and Zwane, T. (2017). Sustainable construction practices: A lazy view of construction professionals in the South Africa construction industry. *The 8th International Conference on Applied Energy Procedia*, 105: 3003–3010

Aigbavboa, C., Oke, A. and Tyali, S. (2016). Unethical practices in South African construction industry. In Emuze, F. (Ed.), *Proceedings of the 5th Construction Management Conference*. Nelson Mandela Metropolitan University, Port Elizabeth, South Africa, 28–29 November, pp. 15–22

Akbiyikli, R., Dikmen, S.U. and Eaton, D. (2009). Sustainability and the Turkish construction cluster: A general overview. *Proceedings of the Construction and Building Research Conference of the Royal Institution of Chartered Surveyors*, University of Cape Town, Cape Town, 10–11 September

Alabi, A.A. (2012). Comparative study of environmental sustainability in building construction in Nigeria and Malaysia. *Journal of Emerging Trends in Economics and Management Sciences*, 3(6): 951–961

Ali, A.S. and Wen, K.H. (2011). Building defects: Possible solutions for poor construction workmanship. *Journal of Building Performance*, 2(1): 59–69

Al-Kharashi, A. and Skitmore, M. (2008). Causes of delays in Saudi Arabia public sector construction projects. *Construction Management and Economics*, 27(1): 3–23

Al-Momani, A.H. (2000). Examining service quality within construction processes. *Technovation*, 20: 643–651

Al-Omari, J. (1992). *Critique of Aspects of Development Theory Using Construction Industry in a Capital-Surplus Developing Economy as an Exemplar*. Published PhD thesis. University of Reading, British Thesis Services, British Library, London

Al-Sanad, S. (2015). Awareness, drivers, actions, and barriers of sustainable construction in Kuwait. *International Conference on Sustainable Design, Engineering and Construction, Procedia Engineering*, 118: 969–983

Alshihre, F., Chinyio, E., Nzekwe-Excel, C. and Daniel, E.I. (2020). Improving clients' satisfaction in construction projects: The case of Saudi Arabia. *Built Environment Project and Asset Management*, 10(5): 709–723

Al-Tmeemy, S.M.H.M., Abdul-Rahman, H. and Harun, Z. (2011). Future criteria for success of building projects in Malaysia. *International Journal of Project Management*, 29(3): 337–348

Ameh, O.J., Soyingbe, A.A. and Odusami, K.T. (2010). Significant factors causing cost overruns in telecommunication projects in Nigeria. *Journal of Construction in Developing Countries*, 15(2): 49–67

Ametepey, O. and Aigbavboa, C. (2014). Practitioners perspectives for the implementation of sustainable construction in Ghana. *Proceedings of the DII-2014 Conference on Infrastructure Investments in Africa*, Livingstone, Zambia, 25–26 September, pp. 114–124

Ametepey, O., Aigbavboa, C. and Ansah, K. (2015). Barriers to successful implementation of sustainable construction in the Ghanaian construction industry. *6th International Conference on Applied Human Factors and Ergonomics and the Affiliated Conferences, Procedia Manufacturing*, 3(1): 1682–1689

Ameyaw, E.E., Pärn, E., Chan, A.P.C., Owusu-Manu, D., Edward, D.J. and Darko, A. (2017). corrupt practices in the construction industry: Survey of Ghanaian experience. *Journal of Management in engineering*, 33(6): 1–11

Amu, O.O. and Adesanya, D.A. (2011). Mathematical expressions for explaining project delays in South western Nigeria. *Singapore Journal of Scientific Research*, 1: 59–67.

Arditi, D. and Gunaydin, H.M. (1997). Total quality management in the construction process. *International Journal of Project Management*, 15(4): 235–243

Ashworth, A. (2004). *Cost Studies of Buildings*, 4th edition. Pearson/Prentice Hall, Harlow, England and New York.

Atkinson, A.A., Waterhouse, J.H. and Wells, R.B. (1997). A stakeholder approach to strategic performance measurement. *Sloan Management Review*; Cambridge, 38(3): 25–37

Atkinson, R. (1999). Project management: Cost, time and quality, two best guesses and a phenomenon, it's time to accept other success criteria. *International Journal of Project Management*, 17(6): 337–342

Attabra-Yartey, B. (2012). *Assessing the Impact of Occupational Health and Safety Needs*. An unpublished thesis submitted to the Department of Managerial Science, Kwame Nkrumah. University of Science and Technology in partial fulfilment of the requirement for the Degree of Master of Business Administration School of Business College of Arts and Social Sciences

Ayarkwa, J., Acheampong, A., Wiafe, F. and Boateng, B.E. (2017). Factors affecting the implementation of sustainable construction in Ghana: The architect's perspective. *ICIDA 2017–6th International Conference on Infrastructure Development in Africa*, Knust, Kumasi, Ghana, 12–14 April, pp. 377–386

Ayedun, C.A. and Oluwatobi, A.O. (2011). Issues and challenges militating against the sustainability of affordable housing provision in Nigeria. *Business Management Dynamics*, 1(4): 1–8

Ayodele, E.O. and Alabi, O.M. (2011). Abandonment of construction projects in Nigeria: Causes and effects. *Journal of Emerging Trends in Economics and Management Sciences*, 2(2): 142–145

Azis, A.A., Memon, A.H., Rahman, I.A. and Karim, A.T. (2013). Controlling cost overrun factors in construction projects in Malaysia. *Research Journal of Applied Sciences, Engineering and Technology*, 5(8): 2621–2629

Baron, N. and Donath, D. (2016). Learning from Ethiopia – A discussion on sustainable building. *Proceedings of SBE16 Hamburg International Conference on Sustainable Built Environment Strategies – Stakeholders – Success factors*, Hamburg, Germany, 7–11 March

BBC News. (2012). Nigerians living in poverty rise to nearly 61%, 2012. Available at: www.bbc.com/news/world-africa-17015873 [accessed 19–8–20219]

Beheiry, S. (2006). *Measuring Sustainability*. The American University of Sharjah, Sharjah. Available at: www.scheller.gatech.edu/centers [accessed 19–8–2019]

Bennett, J. (1985). *Construction Project Management*. Butterworths, London

Bertelsen, S. and Koskela, L. (2004). Construction beyond lean: A new understanding of construction management. *Proceedings of the International Group Lean Conference*. Helsingør, Denmark, p. 12

Brundtland Report. (1987). *Our Common Future*. United Nations Assembly, Report of the World Commission on Environment and Development. Annex to General Assembly Document A/42/427

Bubshait, A.A. and Almohawis, S.A. (1994). Evaluating the general conditions of a construction contract. *International Journal of Project Management*, 12(3): 133–135

Bust, P.D., Gibb, A.G.F. and Pink, S. (2008). Managing construction health and safety: Migrant workers and communicating safety messages. *Safety Science*, 46(4): 585–602

Bustani, S.A. (2000). Availability and quality of construction craftsmen and artisans in the Nigerian construction industry. *Journal of Construction Technology and Management*, 3(1): 91–103

Chan, A.P.C. (2001). Framework for measuring success of construction projects. *Report 2001–003C-01 School of Construction Management and Property Queensland University of Technology Brisbane*, Australia

Chan, A.P.C. and Chan, A.P.C. (2004). Key performance indicators for measuring construction project success. *Benchmarking*, 11(2): 203–221

Chan, A.P.C., Scott, D. and Chan, A.P.C. (2004). Factors affecting the success of a construction project. *Journal of Construction Engineering and Management*, 130(1): 153–155

Chan, A.P.C., Scott, D. and Lam, E.W. (2002). Framework of success criteria for design/build projects. *Journal of Management in Engineering*, 18(3): 120–128

Chan, D.W.M. and Kumaraswamy, M.M. (1997). A comparative study of causes of time overruns in Hong Kong construction projects. *International Journal of Project Management*, 15(1): 55–63

Cheelo, C. and Liebenthal, R. (2018). The role of the construction sector in influencing natural resource use, structural change, and industrial development in Zambia. WIDER Working Paper 2018/172. Available at: www.wider.unu.edu/sites/default/files/Publications/Working-paper/PDF/wp2018-172.pdf [accessed 24–11–2020]

Chilipunde, R.L. (2010). *Constraints and Challenges Faced by Small, Medium and Micro Enterprise Contractors in Malawi.* A treatise submitted to the Faculty of Engineering, the Built Environment and Information Technology, Nelson Mandela Metropolitan University, School of the Built Environment

Chinedu, C.N. and Fidelis, I.E. (2011). Building construction project management success as a critical issue in real estate development and investment. *American Journal of Social and Management Sciences,* 2(1): 56–75

Chiocha, C., Smallwood, J. and Emuze, F. (2011). Health and safety in the Malawian construction industry. *Journal of Acta Structilia,* 18(1): 68–80

Chovichien, V. and Nguyen, T.A. (2013). List of indicators and criteria for evaluating construction project success and their weight assignment. *4th International Conference on Engineering Project and Production,* Bangkok, Thailand, 23–25 October.

Chris, H. (1998). *Project Management for Construction,* 1st edition. Prentice-Hall, Hoboken, NJ

Construction Industry Development Board (CIDB). (2007). The building and construction materials sector, challenges and opportunities. Available at: www.cidb.org.za/publications/Documents/The%20Building%20and%20Construction%20Materials%20Sector,%20Challenges%20and%20Opportunities.pdf

Construction Industry Development Board (CIDB). (2011). *Construction Quality in South Africa: A Client Perspective.* CIDB, Pretoria. Available at: www.cidb.org.za/Documents/KC/cidb_ Publications/Ind_Reps_Other/Construction_Quality_in_SA_Client_Perspective_2010_06_29_final.pdf [accessed 24–11–2020]

Construction Industry Development Board (CIDB). (2017). *Construction Monitor; Employment.* Third Quarter, CIDB Pretoria, South Africa, October, pp. 1–20

Construction Industry Development Board (CIDB). (2018). Construction monitor – employment Q3. Available at: www.cidb.org.za/publications/Documents/Construction%20Monitor%20-%20October%202018.pdf [accessed 24–11–2020]

Department of Statistics Malaysia. (2020). Malaysia economic performance second quarter 2020. Department of statistics Malaysia official portal. Available at: www.dosm.gov.my/v1/index.php?r=column/cthemeByCatandcat=100andbul_id=NS9TNE9yeHJ1eHB6cHV1aXBNQlNUZz09andmenu_id=TE5CRUZCblh4ZTZMODZIbmk2aWRRQT09 [accessed 24–11–2020]

De Soto, B.G., Streule, T., Klippel, M., Bartlome, O. and Adey, B.T. (2018). Improving the planning and design phases of construction projects by using a case-based digital building system. *International Journal of Construction Management:* 1–12

Ding, G. and Langston, C. (2004). Multiple criteria sustainability modelling: Case study on school buildings. *International Journal of Construction Management,* 4(2): 13–26

Dixit, S., Pandey, A.K., Mandal, S.T. and Bansal, S. (2017). A study of enabling factors affecting construction productivity: Indian scenario. *International Journal of Civil Engineering and Technology,* 8(6): 741–758

Djokoto, S.D., Dadzie, J. and Ohemeng-Ababio, E. (2014). Barriers to sustainable construction in the Ghanaian construction industry: Consultants' perspectives. *Journal of Sustainable Development,* 7(1): 134–143

Dozzi, P., Hartman, F., Tidsbury, N. and Ashrafi, R. (1996). More stable owner-contractor relationships. *Journal of Construction Engineering & Management,* 122(1): 30–35

Durdyev, S. and Ismail, S. (2012). Pareto analysis of on-site productivity constraints and improvement techniques in construction industry. *Scientific Research and Essays,* 7(4): 824–833

Elkhalifa, A.A. (2012). *The Construction and Building Materials Industries for Sustainable Development in Developing Countries: Appropriate and Innovative Building Materials and Technologies for Housing in Sudan.* PhD thesis, University of Camerino, Italy

Elkhalifa, A.A. (2016). The magnitude of barriers facing the development of the construction and building materials industries in developing countries, with special reference to Sudan in Africa. *Habitat International*, 54: 189–198

Endut, I.R., Akintoye, A. and Kelly, J. (2009). Cost and time overruns of projects in Malaysia. Available at: www.irbnet.de/daten/iconda/CIB10633.pdf

Enshassi, A., Mohamed, S. and Abushaban, S. (2009). Factors affecting the performance of construction projects in the Gaza strip. Gaza Palestine. *Journal of Civil Engineering and Management*, 15(3): 269–280

Fadhi, M.D. and Tan, F.H. (2001). Developing world class construction companies in Singapore. *Construction Management and Economics*, 19(6): 591–599

Famakin, I.O. and Ojo, A. (2013). Satisfaction of public clients' patronage of construction contractors in Nigeria. *Journal of Architecture, Planning and Construction Management*, 2(2): 37–49

Ferreira, E.J. and Van Loggerenberg, N.F.J. (2011). Perceptions on safety management within South African small and medium enterprises (SMEs): Research and theory. *Africa Safety Promotion Journal*, 9(2): 25–42

Flyvbjerg, B., Holm, M. and Buhl, S. (2003). How common and how large are cost overruns in transport infrastructure projects? *Transport Reviews*, 23(1): 71–88

Folorunso, T.A. and Awodele, O.A. (2015). Assessment of client's needs and satisfaction at various stages of building projects delivery process in Lagos State. In Ogunsemi, D.R., Awodele, O.A. and Oke, A.E. (Eds.), *Proceedings of the 2nd Nigerian Institute of Quantity Surveyors Research Conference*. Federal University of Technology, Akure

Fourie, E.S. (2008). Non-standard workers: The South African context, international law and regulation by the European Union. *PER/PELJ*, 11(4): 1–44

Fox, P.W. (2003). *Construction Industry Development: Analysis and Synthesis of Contributing Factors*. An unpublished Thesis submitted for the degree of Doctor of Philosophy to the School of Construction Management and Property Faculty of Built Environment and Engineering Queensland University of Technology, Queensland, Australia

Frimpong, Y., Oluwoye, J. and Crawford, L. (2003). Causes of delay and cost overruns in construction of groundwater projects in developing countries; Ghana as a case study. *International Journal of Project Management*, 21: 321–326

George, I.N. and Ukpong, D.E. (2013). Contemporary social problems in Nigeria and its impact on national development: Implication for guidance and counselling services. *Journal of Educational and Social Research*, 3(2): 167–173

Giusca, R., Corobceanu, V. and Nenov, M.C. (2009). Study on the concept of quality in construction works. *Journal of Applied Sciences*, 9: 2778–2785

Gould, F.E. and Joyce, N.C. (2008). *Construction Project Management*, 3rd edition. Prentice-Hall, Upper Saddle River, NJ

Gray, P.C.R. and Wiedemann, P.M. (1999). Risk management and sustainable development: Mutual lessons from approaches to the use of indicators. *Journal of Risk Research*, 2(3): 201–218

Griffith, A., Stephenson, P. and Bhutto, K. (2005). An integrated management system for construction quality, safety and environment: A framework for IMS. *International Journal of Construction Management*, 5(2): 51–60

Haseeb, M., Xinhai-Lu, Aneesa Bibi, A., Maloof-ud-Dyian, and Rabbani, W. (2011). Causes and effects of delays in large construction projects of Pakistan. *Kuwait Chapter of Arabian Journal of Business and Management Review*, 1(4): 18–42

Haupt, T.C. and Harinarain, N. (2016). The image of the construction industry and its employment attractiveness. *Acta Structilia*, 23(2): 79–108

Hendricks, B. (2016). *Guideline for Sustainable Building. Future-Proof Design, Construction and Operations of Buildings.* Federal Ministry for the Environment, Nature Conservation, Building and Nuclear Safety, Berlin, Germany

Heravi, G. and Ilbeigi, M. (2012). Development of a comprehensive model for construction project success evaluation by contractors. *Engineering, Construction and Architectural Management,* 19(5): 526–542

Ho, C.M. (2011). Ethics management for the construction industry. *Engineering, Construction and Architectural Management,* 18(5): 516–537

Human Science Research Council (HSRC). (2009). *Skills Shortages in South Africa: Case Studies of Key Professions.* Education and Skills Development. Available at: www.hsrc press.ac.za/product.php?productid=2257andcat=1andpage=2 [accessed 19–11–2019]

Hussin, J.M., Rahman, I.A. and Memon, A.H. (2013). The way forward in sustainable construction: Issues and challenges. *International Journal of Advances in Applied Sciences,* 2(1): 15–24

Ibrahim, R., Roy, M.H., Ahmed, Z.U. and Imtiaz, G. (2010). Analyzing the dynamics of the global construction industry: Past, present and future. *Benchmarking: An International Journal,* 17(2): 232–252

Idoro, G.I. (2009). Evaluating levels of project planning and their effects on performance in the Nigerian construction industry. *The Australasian Journal of Construction Economics and Building,* 9(2): 39–50

International Labour Organization. (1998). *Low-Cost Ways of Improving Working Conditions, 100 Examples from Asia.* ILO, Asia. Available at: http://collections.infocollections.org/ ukedu/uk/d/Jh2343e/ [accessed 23–6–2018]

International Labor Organization (ILO). (2001). *The Construction Industry in the Twenty-First Century: Its Image, Employment Prospects and Skill Requirements.* ILO, Geneva. Available at: www.ilo. org/public/english/standards/relm/gb/docs/gb283/pdf/tmcitr.pdf [accessed 23–6–2018]

Isa, M., Rahman, M.M.G.M.A., Sipan, I. and Hwa, T.K. (2013). Factors affecting green office building investment in Malaysia. *Procedia – Social and Behavioral Sciences,* 105: 138–148

Iyer, K.C. and Jha, K.N. (2005). Factors affecting cost performance: Evidence from Indian construction projects. *International Journal of Project Management,* 23: 283–295

Jackson, S. (2002). Project cost overruns and risk management. In Greenwood, D. (Ed.), *Proceedings 18th Annual Conference of the Association of Researchers in Construction Management,* ARCOM, University of Northumbria, Newcastle, September

Kadiri, Z.O., Nden, T., Avre, G.K., Oladipo, T.O., Edom, A., Samuel, P.O. and Ananso, G.N. (2014). Causes and effects of accidents on construction sites (a case study of some selected construction firms in Abuja F.C.T Nigeria). *IOSR Journal of Mechanical and Civil Engineering,* 11(5): 66–72

Kaliba, C., Muya, M. and Mumba, K. (2009). Cost escalation and schedule delays in road construction projects in Zambia. *International Journal of Project Management,* 27: 522–531

Kam, C.W. and Tang, S.L. (1998). ISO 9000 for building and civil engineering contractors, *Journal of Hong Kong Institution of Engineering,* 5(2): 6–10

Karna, S. and Jonnonen, J. (2005). Project feedback as a tool for learning. A Paper Presented at IGLC-13, Sydney

Karna, S., Junnonen, J. and Kankainen, J. (2004). Customer satisfaction in construction. Available at: www.iglcstorage.blob.core.windows.net [accessed 23–6–2018]

Kartam, N.A., Flood, I. and Koushki, P. (2000). Construction safety in Kuwait: Issues, procedures, problems, and recommendations. *Safety Science,* 36(3): 163–184

Kazaz, A. and Birgonul, M.T. (2005). Determination of quality level in mass housing projects in Turkey. *Journal of Construction Engineering and Management*, 131(2): 195–202

Kazaz, A. and Ulubeyli, S. (2009). Strategic management practices in Turkish construction firms. *Journal of Management in Engineering*, 25(4): 185–194

Khalfan, M.M.A. and Anumba, C.J. (2000). Implementation of concurrent engineering in construction – readiness assessment. *Proceedings of Construction Information Technology (CIT2000) Conference*, Reykjavik, Iceland, 28–30 June, Vol. 1, pp. 544–555

Kheni, N., Dainty, A.R.J. and Gibb, A.G.F. (2008). Health and safety management in developing countries: A study of construction SME's in Ghana. *Journal of Construction Management and Economics*, 26(11): 1159–1169

Khosravi, S. and Afshari, H. (2011). A success measurement model for construction projects. *International Conference on Financial Management and Economics IPEDR*, 11: 186–190

Kibert, C.J. (2013). *Sustainable Construction – Green Building Design and Delivery*. John Wiley and Sons, NJ

Kikwasi, G.J. (2012). Causes and effects of delays and disruptions in construction projects in Tanzania. *Australasian Journal of Construction Economics and Building, Conference Series*, 1(2): 52–59

Koushki, P.A., Al-Rashid, K. and Kartam, N. (2005). Delays and cost increases in the construction of private residential projects in Kuwait. *Construction Management and Economics*, 23(3): 285–294

Laryea, S. and Mensah, S. (2010). The evolution of indigenous contractors in Ghana. In Laryea, S., Leiringer, R. and Hughes, W. (Eds.), *Proceedings West Africa Built Environment Research Conference, WABER*, Accra, Ghana, 27–28 July, pp. 579–588.

Lee, J.K. (2008). Cost overrun and cause in Korean social overhead capital projects: Roads, Rails, Airports, and ports. *Journal of Urban Planning and Development*, 134(2): 59–62

Le-Hoai, L., Lee, Y.D. and Lee, J.Y. (2008). Delay and cost overruns in Vietnam large construction projects: A comparison with other selected countries. *KSCE Journal of Civil Engineering*, 12: 367–377

Leibing, R. (2001). *The Construction Industry: Process Players*. Prentice Hall, Upper Saddle River, NJ

Lim, C.S. and Mohamed, M.Z. (1999). Criteria of project success: An exploratory re-examination. *International Journal of Project Management*, 17(4): 243–248

Loosemore, M., Lingard, H., Walker, D.H.T. and Mackenzie, J. (1999). Benchmarking safety management systems in contracting organisations against best practice in other industries. In Singh, A., Hinze, J.H. and Coble, R.J. (Eds.), *Implementation of Safety and Health on Construction Sites*. A.A. Balkema, Rotterdam

Lopes, R.P., Dillenburg, S.R. and Schultz, C.L. (2011). Geological and environmental evolution of lagoon system III in the southernmost coastal plain of Rio Grande do Sul state. In *Congresso Da Associação Brasileira De Estudos Do Quaternário, 13*. Resumos Expandidos, Armação de Búzios, p. 6

Love, P.E.D. and Irani, Z. (2003). A project management quality cost information system for the construction industry. *Information & Management*, 40(7): 649–661

Lowe, D.J. and Zhou, L. (2003). Economic factors of sustainable construction. *RICS COBRA Foundation Construction and Building Research Conference*, University of Wolverhampton and The RICS Foundation, London, 1–2 September, pp. 113–126

Makhene, D. and Twala, W.D. (2009). Skilled labour shortages in construction contractors: A literature review. *CIDB (Construction Industry Development Board), 6th Postgraduate Conference on Construction Industry Development*, CIDB, Pretoria/Johannesburg, South Africa, 6–8 September, pp. 128–136

Maseko, C. (2017). Literature on theory and practice on unethical practices in the construction of projects: A case of an emerging economy. *Risk Governance and Control: Financial Markets and Institutions*, 7(4): 214–224

Mashwama, N.X. and Musonda, I. (2014). An investigation on the impact of subcontracting system on the eventual quality of construction facilities in Swaziland – an exploratory study. In Emuze, F.A. and Aigbavboa, C.A. (Eds.), *Conference Proceedings: TG59 People in Construction Conference*, Nelson Mandela Metropolitan University, Port Elizabeth, South Africa, 6–8 April, pp. 191–200

Mbachu, J.I.C. and Nkando, R.N. (2004). Reducing building construction costs: The views of consultants and contractors. *RICS COBRA Research Conference*, Leeds Metropolitan University, UK, 7–8 September

Mbachu, J.I.C. and Nkado, R.N. (2006). Conceptual framework for assessment of client needs and satisfaction in the building development process. *Construction Management and Economics*, 24: 31–44

Mbamali, I. and Okotie, A.J. (2012). An assessment of the threats and opportunities of globalization on building practice in Nigeria. *American International Journal of Contemporary Research*, 4: 143–150

Mbande, C. (2010). Overcoming construction constraints through infrastructure deliver. *Proceedings: The Association of Schools of Construction of Southern Africa (ASOCSA), Fifth Built Environment Conference*, Durban, South Africa, 18–20 July

Memon, A.H. (2013). *Structural Modelling of Cost Overrun Factors in Construction Industry*. An unpublished PhD thesis submitted to the Faculty of Civil and Environmental Engineering University Tun Hussein Onn Malaysia

Memon, A.H., Rahman, I.A., Abdullah, M.R. and Azis, A.A. (2012). The cause factors of large project's cost overrun: A survey in the southern part of peninsular. *International Journal of Real Estate Studies*, 7(2): 1–15

Mengistu, D.G. and Mahesh, G. (2020). Challenges in developing the Ethiopian construction industry. *African Journal of Science, Technology, Innovation and Development*, 12(4): 373–384

Milford, R. (2009). Construction industry development in developing countries; Lessons and opportunities. *3rd International Conference on Concrete and Development*, Tehran, Iran, 27 April

Mousa, A.A. (2015). Business approach for transformation to sustainable construction: An implementation on a developing country. *Resources, Conservation and Recycling*, 101: 9–19

Muhammad, M. (2012), Poverty and national security: A review of policies 1960–2010. *JORIND*, 10(2): 243–251

Muiruri, G. and Mulinge, C. (2014). Health and safety on construction project sites in Kenya: A case study of construction projects in Nairobi Country. *In FIG Congress 2014: Engaging Challenges-Enhancing the relevance*, Kuala Lumpur, Malaysia

Mulalo, R., Ibrahimu, K. and Nwobodo-Anyadiegwu, E. (2018). Project cost overrun in the South African construction sector: A case study of Johannesburg Metropolis. *Proceedings of the International Conference on Industrial Engineering and Operations Management*, Pretoria/Johannesburg, South Africa, 29 October–1 November

Nagarajan, D. (2018). Challenges in construction industry in India – 2018. Available at: www.zerektech.com/challenges-in-construction-industry-in-india-2018/ [accessed 19–11–2019]

Nam, C.H. and Tatum, C.B. (1988). Major characteristic of constructed products and resulting limitations of construction technology. *Construction Management and Economics*, 6: 133–148

National Bureau of Statistics. (2020). *Nigerian Gross Domestic Product Report (Q3)*. National Bureau of Statistics PDF. Available at: www.nigerianstat.gov.ng/ [accessed 24–11–2020]

Navon, R. (2005). Automated project performance control of construction projects. *Automation in Construction*, 14: 467–476

Nel, P.S. and Werner, A. (2017). *Human Resource Management*, 10th edition. Oxford University Press, CapeTown, South Africa

Nhabinde, V., Marrengula, C.P. and Ubisse, A. (2012). The challenges and the way forward for the construction industry in Mozambique. *A Report of the International Growth Centre Mozambique*. Available at: www.theigc.org/publication/the-challenges-and-the-way-forward-for-the-construction-industry-in-mozambique-working-paper/ [accessed 24–11–2020]

Noulmanee, A., Wachirathamrojn, J., Tantichattanont, P. and Sittivijan, P. (1999). Internal causes of delays in highway construction projects in Thailand. Available at: www.ait.clet.com [accessed 24–11–2020]

Nzekwe-Excel, C. (2007). Improved client satisfaction: A strategic. Approach in the construction sector. *3rd Scottish Conference for Postgraduate Researchers of the Built and Natural Environment*, Glasgow Caledonian University. Available at: www.irbnet.de/daten/iconda/CIB10781.pdf [accessed 19–11–2019]

Obiegbu, M.E. (2003). Building development process – a search for cohesiveness and teamwork. A paper presented at the Reception Ceremony of the 13th NIOB President by the Anambra State Chapter, November

Odediran, S.J., Adeyinka, B.F., Opatunji, O.A. and Morakinyo, K.O. (2012). Business structure of indigenous firms in the Nigerian construction industry. *International Journal of Business Research and Management*, 3(5): 255–264

Odeyinka, H.A. and Yusif, A. (1997). The causes and effect of construction delays on completion cost of housing projects in Nigeria. *Journal of Financial Management Property and Construction*, 2(3): 31–44

Ofori, G. (1980). *The Construction Industries of Developing Countries: The Applicability of Existing Theories and Strategies for Their Improvement and Lessons for the Future- the Case of Ghana*. Unpublished PhD thesis, University College, London

Ofori, G. (2000). Globalization and construction industry development: Research opportunities. *Construction Management and Economics*, 18: 257–262

Ofori, G. (2018). Construction in developing countries: Need for new concepts. *Journal of Construction in Developing Countries*, 23(2): 1–6

Ogbu, C.P. (2017). Survival practices of indigenous construction firms in Nigeria. *International Journal of Construction Management*, 18(1): 78–91

Ogunlana, S., Promkuntong, K. and Jearkjirm, V. (1996). Construction delays in a fast-growing economy: Comparing Thailand with other economies. *International Journal of Project Management*, 14(1): 37–45

Ogunmankinde, O.E., Sher, W. and Maund, K. (2019). Challenges of the Nigerian construction industry: A systematic review. In *IISBE Forum of Young Researchers in Sustainable Building*. Czech Republic, Prague, 1 July

Ogunsemi, D.R. (2015). Value for money in construction projects: The quantity surveyor's quest. *71st Inaugural Lecture delivered by Prof. D. R. Ogunsemi at the 2500 Capacity Auditorium*, Federal University of Technology, Akure, Ondo State

Ogunsemi, D.R. and Jagboro, G.O. (2006). Time-cost model for building projects in Nigeria. *Construction Management and Economics*, 24: 253–258

Oke, A.E., Aghimien, D.O., Aigbavboa, C.O. and Koloko, N. (2018). Challenges of digital collaboration in the South African construction industry. *Proceedings of the International*

Conference on Industrial Engineering and Operations Management, Bandung, Indonesia, 6–8 March, pp. 2472–2482

Oke, A.E., Aghimien, D.O. and Olatunji, S.O. (2015). Implementation of value management as an economic sustainability tool for building construction in Nigeria. *International Journal for Managing Value and Supply Chain*, 6(4): 55–64

Oke, A.E. and Aigbavboa, C.O. (2017). *Sustainable Value Management for Construction Projects.* Springer International Publishing AG, Cham, Switzerland

Okhiria, A.O. and Obadeyi, T. (2015). Poverty, an African epidemic: Empirical evidence of Nigeria. *Developing Country Studies*, 5(6): 29–39

Okoye, P.U., Okolie, K.C. and Ngwu, C. (2017). Multilevel safety intervention implementation strategies for Nigeria construction industry. *Journal of Construction Engineering*: 84–96

Oladinrin, T.O., Ogunsemi, D.R. and Aje, I.O. (2012). Role of construction sector in economic growth: Empirical evidence from Nigeria. *FUTY Journal of the Environment*, 7(1): 50–60

Olatunji, O.A. (2008). A comparative analysis of tender sums and final costs of public construction and supply projects in Nigeria. *Journal of Financial Management of Property and Construction*, 13(1): 60–79

Olatunji, S.O., Oke, A.E., Aghimien, D.O. and Ogunwoye, O.S. (2016a). Implementation of code of ethics among quantity surveying firms in Nigeria. *Developing Country Studies*, 6(5): 71–76

Olatunji, S.O., Oke, A.E., Aghimien, D.O. and Saidu, S.A. (2016b). Effect of construction project performance on economic development of Nigeria. *Journal of Economics and Sustainable Development*, 7(12): 142–149.

Omonori, A. and Lawal, A. (2014). Understanding customers' satisfaction in construction industry in Nigeria. *Journal of Economics and Sustainable Development*, 5(25): 115–120

Omoregie, A. and Radford, D. (2006). Infrastructure delays and cost escalation: Causes and effects in Nigeria. *Proceeding of Sixth International Postgraduate Research Conference*, Delft University of Technology, Delft, Netherland, pp. 79–93

Opoku, A. and Ahmed, V. (2014). Embracing sustainability practices in UK construction organizations: Factors facing intra-organizational leadership. *Built Environment Project and Asset Management*, 4(1): 90–107

Orji, E., Enebe, E.C. and Onoh, F.E. (2016). Accidents in building construction sites in Nigeria; a case of Enugu state. *International Journal of Innovative Research*, 5(4): 244–248

Osaily, N.Z. (2010). *The Key Barriers to Implementing Sustainable Construction in West Bank – Palestine.* Robert Kennedy College and Zurich, University of Wales, Cardiff, United Kingdom

Oswald, D., Sherratt, F. and Smith, S. (2013). Exploring factors affecting unsafe behaviours in construction. In Smith, S.D and Ahiaga-Dagbui, D.D (Eds.), *Proceedings 29th Annual ARCOM Conference.* Association of Researchers in Construction Management, Reading, UK, 2–4 September, pp. 335–344

Othman, I., Shafiq, N. and Nuruddin, M.F. (2017). Quality planning in construction project. *IOP Conference Series: Materials Science and Engineering*, 291: 1–5

Oyewobi, L.O., Ganiyu, B.O., Oke, A.A., Ola-Awo, A.W. and Shittu, A.A. (2011). Determinants of unethical performance in Nigerian construction industry. *Journal of Sustainable Development*, 4(4): 175–182

Oyedele, L.O., Jaiyeoba, B.E., Kadiri, K.O., Folagbade, S.O., Tijani, I.K. and Salami, R.O. (2015). Critical factors affecting construction quality in Nigeria: Evidence from industry professionals. *International Journal of Sustainable Building Technology and Urban Development*, 6(2): 103–113

Ozmec, M.N., Karlsen, I.L., Kines, P., Andersen, L.P.S. and Nielsen, K.J. (2015). Negotiating safety practice in small construction companies. *Safety Science*, 71: 275–281

Potty, N.S., Idrus, A.B. and Ramanathan, C.T. (2011). Case study and survey on time and cost overrun of multiple D&B projects. *IEEE 2011 National Postgraduate Conference*, Kuala Lumpur, pp. 1–6

Powmya, A. and Abidin, Z.N. (2014). The challenges of green construction in Oman. *International Journal of Sustainable Construction Engineering and Technology*, 5(1): 33–41

Puplampu, B.B. and Quartey, S.H. (2012). Key issues on occupational health and safety practices in Ghana. *International Journal of Business and Social Sciences*, 3: 151–156

Rameezdeen, R. (2007). Image of the construction industry. In Sexton, M., Kähkönen, K. and Shu-Ling Lu (Eds.), *Revaluing Construction: A W065 Organisation and Management of Construction Perspective*. CIB Publication, The Netherlands, Vol. 313, pp. 76–87

Rogerson, R.J. (1999). Quality of life and city competitiveness. *Urban Studies*, 36(5–6): 969–985

Saka, N. and Lowe, J. (2010). An assessment of linkages between the construction sector and other sectors of the Nigerian economy. *The Construction, Building and Real Estate Research Conference of the Royal Institution of Chartered Surveyors*, Held at Dauphine Université, Paris, 2–3 September

Salter, A. and Torbett, R. (2003). Innovation and performance in engineering design. *Journal of Construction Management and Economics*, 21: 573–580

Sambasivan, M. and Soon, Y.W. (2007). Causes and effects of delays in Malaysian construction industry. *International Journal of Project Management*, 25(5): 517–526

Sanchez, P., Swaminathan, M.S., Dobie, P. and Yuksel, N. (2005). Halving hunger: It can be done. In *UN Millennium Project, Summary Version of the Report of the Task Force on Hunger*. The Earth Institute at Columbia University, New York

Sawhney, A., Agnihotri, R. and Kumar Paul, V. (2014). Grand challenges for the Indian construction industry. *Built Environment Project and Asset Management*, 4(4): 317–334

Serpell, A. and Alarco´n, L.F. (1998). Construction process improvement methodology for construction projects. *International Journal of Project Management*, 16(4): 215–221

Shah, R.K. and Alotaibi, M. (2017). A study of unethical practices in the construction industry and potential preventive measures. *Journal of Advanced College of Engineering and Management*, 3: 55–77

Shenhar, A.J., Levy, O. and Dvir, D. (1997). Mapping the dimensions of project. *Success. Project Management Journal*: 5–13

Silva, G.A.S.K., Warnakulasuriya, B.N.F. and Arachchige, B.J.N (2016). Criteria for construction project success: A literature review. *13th International Conference on Business Management*, Colombo, Sri Lanka, 8 December

Singh, R.K., Murty, H.R., Gupta, S.K. and Dikshit, A.K. (2009). An overview of sustainability assessment methodologies. *Ecological Indicators*, 9: 189–212

Sjostrom, C. and Bakens, W. (1999). Sustainable construction: Why, how and what. *Building Research and Information*, 27(6): 347–353

Smallwood, J.J. (2004). Optimum cost: The role of health and safety. In Verster, J.J.P. (Ed.), *International Cost Engineering Council 4th World Congress*. Cape Town, 17–21 April.

Spellman, F.R. (2016). *Occupational Safety and Health Simplified for the Industrial Workplace*. Bernan Press, MD

Statista Research Department. (2019). *China's Construction Industry Contribution Share to GDP 2018 to 2021*. Statista Research Department. Available at: www.statista.com/statistics/1068213/china-construction-industry-gdp-contribution-share/#:~:text=China's%20construction%20industry%20contribution%20share%20to%20

GDP%202018%20to%202021andtext=In%202018%2C%20the%20construction%20 industry,a%20five%20percent%20annual%20growth [accessed 24–11–2020]

Stats SA. (2017). Poverty on the rise in South Africa. Available at: www.statssa.gov. za/?p=10334

Sunindijo, R.Y. (2015). Improving safety among small organisations in the construction industry: Key barriers and improvement strategies. *Procedia Engineering*, 125: 109–116

Takim, R. and Adnan, H. (2008). Analysis of effectiveness measures of construction project success in Malaysia. *Asian Social Science*, 4(7): 74–91

Takim, R. and Akintoye, A. (2002). Performance indicators for successful construction project performance. In Greenwood, D. (Ed.), *18th Annual Association of Researchers in Construction Management Conference*, ARCOM, University of Northumbria, Newcastle, pp. 545–555.

Torbica, Z.M. and Stroh, R.C. (2001). Customer satisfaction in home building. *Journal of Construction Engineering and Management*, 127(1): 82–86

Transparency International. (2019). Corruption perceptions index 2018. Available at: www. transparency.org/cpi2018

Tucker, R.L., Haas, C.T., Glover, R.W., Alemany, C.H., Carley, L.A., Eickmann, J.A., Rodriguez, A.M. and Shields, D. (1999). Craft workers' experiences with and attitudes towards multiskilling. In *Report No. 3, Center for Construction Industry Studies*. The University of Texas at Austin, Austin, TX

Ubani, K.A., Okorocha, S.C. and Emeribe, I. (2013). Analysis of factors influencing time and cost overruns on construction projects in South Eastern Nigeria. *International Journal of Management Sciences and Business Research*, 2(2): 73–84

United State Census Bureau. (2002). *All Other Speciality Trade Contractors: 2002 Economic Census*. Economics and Statistics Administration, US Department of Commerce, Washington, DC

van Wyk, L. (2004). *A Review of the South African Construction Industry Part 2: Sustainable Construction Activities*. CSIR Boutek, Pretoria, South Africa. Available at: www.csir. co.za/akani

Vee, C. and Skitmore, R. (2003). Professional ethics in the construction industry. *Engineering Construction and Architectural Management*, 10(2): 117–127

Vitharana, V.H.P., De Silva, G.H.M.J. and De Silva, S. (2015). Health hazards, risk and safety practices in construction sites – a review study. *Engineer: Journal of the Institution of Engineers*, Sri Lanka, 48(3): 35–44

Wadick, P. (2010). Safety culture among subcontractors in the domestic housing construction industry. *Journal Structural Survey*, 28(2): 108–120

Wells, J. (1985). The role of construction in economics growth and development. *Habitat International*, 9(1): 55–70

Wetangula, J. and Mazurewicz, M. (2017). *Market Research Preliminary Report: The Construction Market in Kenya*. Polish Investment and Trade Agency – Nairobi Office. Available at: www.paih.gov.pl

Windapo, A.O. and Cattell, K. (2010). Perceptions of key construction and development challenges facing the construction industry in South Africa. *The Association of Schools of Construction of Southern Africa (ASOCSA), Fifth Built Environment Conference*, Durban, South Africa, 18–20 July

Windapo, A.O. and Cattell, K. (2013). The South African construction industry: Perceptions of key challenges facing its performance, development and growth. *Journal of Construction in Developing Countries*, 18(2): 65–79

Wold, J., Lædre, O. and Lohne, J. (2019). Questionable practice in the processing of building permits in Norway. In Pasquire, C. and Hamzeh, F.R. (Eds.), *Proceeding 27th Annual*

Conference of the International Group for Lean Construction, IGLC, Dublin, Ireland, pp. 1151–1162

World Economic Forum. (2016). Shaping the future of construction. A breakthrough in mindset and technology. Available at: http://www3.weforum.org/docs/WEF_Shaping_ the_Future_of_Construction_full_report__.pdf [accessed 20–8–2018]

Yan, H. and Damian, P. (2008). Benefits and barriers of building information modelling. In Ren, A., Ma, Z. and Lu, X (Eds.), *Proceedings of the 12th International Conference on Computing in Civil and Building Engineering*, Tingshua University Press, Beijing, China, 16–18 October

Zawdie, G. and Langford, D.A. (2002). Influence of construction-based infrastructure on the development process in Sub-Saharan Africa. *Building Research and Information*, 3(3): 160–170

Zingoni, T. (2020). Deconstructing South Africa's construction industry performance. *Mail and Guardian*. Available at: https://mg.co.za/opinion/2020-10-19-deconstructing-south-africas-construction-industry-performance/#:~:text=The%20South%20Afri can%20construction%20sector,for%20construction%20in%20the%20country [accessed 24–11–2020]

Zujo, V. and Car-Pusic, D. (2008). Application of "time-cost" model in construction project management. *Proceedings, 8th International Conference on Organization, Technology and Management in Construction*, Umag, Croatian Association for Organization in Construction, Zagreb, pp. 1–7

Zujo, V., Car-Pusic, D. and Brkan-Vejzovic, A. (2010). Contracted price overrun as contracted construction time overrun function. *Technical Gazette*, 17: 23–29

3 The fourth industrial revolution and digitalisation

Introduction

This chapter discusses the fourth industrial revolution (4IR) concept by giving an insight into the evolutional trends in industrial revolutions and clearly showcasing the feature of 4IR that makes it unique compared to the previous industrial revolutions. The applications of 4IR features in different sectors like manufacturing, education, finance, agriculture, and construction are also explored in this chapter. The chapter follows that 4IR is driven by physical, digital, and biological technologies of which the digital and physical technologies play a significant role in the digitalisation of construction activities.

The fourth industrial revolution

The fourth industrial revolution (4IR) journey, also known as Industry 4.0, began with the first industrial revolution that occurred in the late 1700s and mid-1800s. This industrial revolution saw the moving away from the agrarian economy to a more mechanical production economy powered by the introduction of water and steam power engine for the manufacturing of products. Then came the second industrial revolution that featured electrical energy for mass production in the early 1900s. A prominent name that brought about this industrial revolution is Henry Ford, whose production line made mass production of cars for public consumption. This revolution was marked by the advent of electricity and assembly line. The third industrial revolution followed in the late 1960s and featured the use of information technology and electronics for industrial production, which gave rise to automated and optimised production methods. This revolution marked the emergence of semiconductors, mainframe computers, personal computers as well as the internet. The third industrial revolution's computerised features earned it the name computer or digital revolution as it is popularly known. Today, 4IR which is built on the idea of the third industrial revolution has become a name on the lips of every manufacturing and business entity who seeks better production and customer satisfaction (Bienhaus & Haddud, 2018; Crnjac et al., 2017; Schwab, 2017). Figure 3.1 gives a brief view of these industrial revolutions.

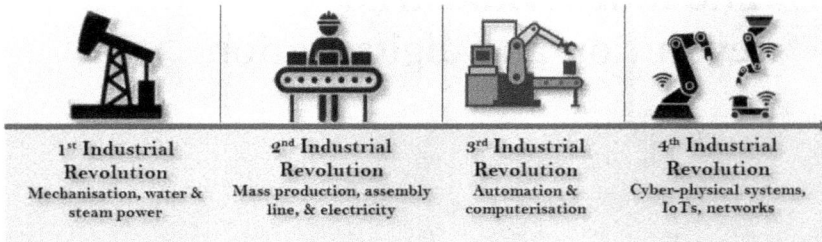

1st Industrial Revolution	2nd Industrial Revolution	3rd Industrial Revolution	4th Industrial Revolution
Mechanisation, water & steam power	Mass production, assembly line, & electricity	Automation & computerisation	Cyber-physical systems, IoTs, networks

Figure 3.1 The industrial revolutions in a snapshot

The term "fourth industrial revolution" was first established at the Hannover Fair in 2011 in Germany. Since then, it has garnered considerable worldwide attention from academics, practitioners, governmental officials, and politicians (Sung, 2018b). The German government presents 4IR as an emerging structure whereby manufacturing and logistics systems employ information and communications networks that are available globally to match automated business processes and production (Bahrin *et al.*, 2016). Through this structure, higher efficiency and change in the traditional production system can be attained, and relationships among participants in the manufacturing system and between humans and machines can improve (Rüßmann *et al.*, 2015). However, it is essential to note that while 4IR can ensure human–machine relationships through smart and connected systems, its scope transcends this basic development. Breakthroughs are being made in gene sequencing, nanotechnology, renewable and reusable materials, and quantum computing. The ability of these technologies to interact with one another and interact across their physical, digital, and biological domains through cyber-physical systems (CPS) makes 4IR a significant advancement from the third industrial revolution (Schwab, 2017).

The term 4IR involves using a wide range of new, emerging, and disruptive digital technologies in the discharge of more productive activities. It is characterised by the ubiquitous and mobile internet, powerful sensors, artificial intelligence (AI), and machine learning (Schwab, 2017). Furthermore, 4IR describes the increased level of automation and digitisation which enables the communication between products and their environment and business partners through creating digital value chain (Dallasega *et al.*, 2018; Oesterreich & Teuteberg, 2016). According to Petrillo *et al.* (2018), 4IR covers three essential aspects: digitisation and increased integration of vertical and horizontal value chains; digitisation of product and service offerings; and innovative digital business models. Prominent features of the 4IR are Internet of Things (IoT), which includes technologies such as radio frequency identification (RFID) sensors that transmit, store, analyse, and process information as well as smartphones that are able to communicate and cooperate with smart components; additive manufacturing (3D printing), which is able to produce complex architectural components through automated manufacturing without extra labour costs; augmented, virtual and mixed realities, which enable augmented and virtual view of objects even before actualising these objects; cloud computing, which provides the opportunity to access integrated services via the internet, with cloud base solutions that allow project stakeholders to access information from any communication device that has access to the internet; big data, which uses intelligent algorithms to collect extensive data, process, and analyse structured and unstructured data; autonomous robots which are used for autonomous production method, as they can execute the works precisely as per specification and are able to function in places restricted to human workers; among others (Celaschi, 2017; Oesterreich & Teuteberg, 2016; Roblek *et al.*, 2016; Stankovic & Neftenov, 2020; Sung, 2018a; Vaidya *et al.*, 2018).

The fourth industrial revolution in diverse sectors

The speed of technological breakthrough is alarming, and this technological disruption is influencing every sector of the economy. Significant evidence abounds in the primary sector, such as agriculture and mining. The secondary sector is also known as the industrial sector, where the raw materials from the primary sector are processed into desired products. The tertiary sector is where the produced goods and services are distributed and consumed. The influence of 4IR in some of these sectors is subsequently discussed.

Manufacturing

The manufacturing industry has taken centre stage in the embrace of the fourth industrial revolution. According to Crnjac *et al.* (2017), Industry 4.0 has become a topical issue in the manufacturing industry due to its focus on creating a smart environment within the production system. Kagermann (2014) in a study of how Industry 4.0 will shape the economy of the future stated that the concept of Industry 4.0 is geared towards creating intelligent products, processes, and procedures through the establishment of smart factories. This factory allows easy communication between workers, machines, and resources. Hermann *et al.* (2016) and Kagermann (2014) noted that Industry 4.0 is mostly IT related to prominent features such as cyber-physical systems, Internet of Things, and cloud computing. This concept creates a smart factory that integrates the cyber-physical systems, monitors physical processes, and creates a virtual copy of the physical world within a modular system, making decentralised decisions.

Industry 4.0 has been described as adopting digital technologies in the manufacturing process to get more agile and flexible products and respond to consumer's demand (Bédard-Maltais, 2017). Vaidya *et al.* (2018) described the Industry 4.0 as a new form of organising and controlling the entire value chain of the life cycle of products with the sole aim of increasing individualised customer requirements. Although the evolution of Industry 4.0 is still in progress (Bienhaus & Haddud, 2018), evidence of a realistic concept can be seen in the areas of the IoT, industrial internet, smart manufacturing, and cloud-based manufacturing

(Vaidya *et al.*, 2018). Bédard-Maltais (2017), while investigating the Canadian manufacturing industry's readiness to adopt the concept of Industry 4.0 observed that about 40% of small and medium manufacturers in the country implemented Industry 4.0 projects. However, out of this 40%, only 3% have fully converted their production process from analogue to digital, while 17% are in the planning phase.

According to Lee (2013), Industry 4.0 is the next phase in digitising the manufacturing sector. This phase results from four strategic features: the massive rise in data, computational power, and connectivity; emerging analytics and business-intelligence capabilities; the emergence of different human–machine interactions; and improvements in the transfer of digital instructions to the physical world. Crnjac *et al.* (2017) further noted that the essential aspects of Industry 4.0 are standardisation which allows connectivity within organisations; management of complex systems which involves developing and applying new models and methods within the activities of the organisation; comprehensive infrastructure such as high-quality information network, and internet connection; security, and privacy to ensure data are protected and intellectual property remains in the possession of the organisation; work organisation and design which involves changing the roles of workers in the organisations by making them more involved in the quest to deliver smart products, and through this, life-long organisation learning is attained; legal framework; and the effective use of resources which involves applying the concept of sustainable manufacturing through saving of raw materials and energy.

To introduce the concept of Industry 4.0 into the manufacturing business and enjoy the benefits thereof, organisations must be willing first to build a digital factory. This can be done by using powerful manufacturing technologies to monitor production and steer it towards the required direction, mine data generated from such products to get insightful ideas that will lead to increased production in the future and reduction in downtime. Sensors can also be employed to monitor and improve the quality of products continuously. Second, such organisations must create digital products through the addition of internet-connected sensors to products in a bid to allow proper monitoring of the product's performance and at the same time provide customers with relevant information relating to the product. The third way digital technologies can be introduced into the construction business is by providing a digital customer experience. This involves using the internet and other advanced digital technologies as avenues to get closer to the customers. A typical example is affording the consumers the ability to customise their orders online and see an order's progress in real time. This can also help in predicting customer's consumption patterns and future wants (Bédard-Maltais, 2017).

Finance

The financial service sector is not spared from the disruption by the fourth industrial revolution. This is evident in the increasing popularity of digital economy – an economy based on digital technologies or economic transactions over the internet. Although the use of technologies offered by Industry 4.0 is believed to still be at its infancy stage in this sector, significant benefits abound from the use of the few technologies that have been adopted (Nair, 2019). Industry 4.0 has seen the rise of new business models and companies known as Financial Technology (FinTech) companies. These companies adopt technological and digital solutions to produce financial products that meet the financial capability of their consumers in a more efficient way (Tanda & Schena, 2019). FinTech can reduce cost while increasing efficiency of financial services; provide flexible and sweatless interaction between customers and their financial providers; and at the same time provide an avenue for financial providers to understand their customers' behaviour and needs which give room for personalisation of services to customers (Centre of Excellence in Financial Services, COEFS, 2017). According to Stefanul (2020), Industry 4.0 is revolutionising the financial services as its impact is evident in online payments, digital loans, plastic money, cryptocurrency, online forex trading, and many other financial activities. This technological disruption is made possible through ubiquitous technology, software development, better internet facilities, and global software teams of qualified professionals.

Ohene-Afoakwa and Nyanhongo (2017) highlighted that Industry 4.0 offers access to the digital world and provides entirely new ways of servicing existing needs. This implies that organisations can have entirely new approaches towards fulfilling their customers' expectations. To enjoy these benefits, the organisation must understand that disruption in the business world is a norm and must be willing to adapt. Klima (2019) has noted that in the era of Industry 4.0, organisations need to be agile across all functions in their organisation so that adaptability, insight, and cost-effective operations can be achieved. Furthermore, organisations must cut back on archaic legacy activities and invest more in research and development and customer experience. It is also time to shift towards robotics and automated processes combined with augmented intelligence and investing in robust, interoperable platforms that will ease using these technologies (Ohene-Afoakwa & Nyanhongo, 2017).

Agriculture

The importance of agriculture to the survival of man cannot be overemphasised. Currently, only about 5% of the world's population is involved in agriculture. This is a significant reduction compared to the over 90% of the world's population involved in agriculture about 200 years ago – the major culprit of this reduction is the different stages of industrialisation that have been experienced. To ensure the continued existence of this vital sector – one that contributes almost 60% of the world's business – developed countries worldwide have promoted the use of mechanisation, automation, and modernisation. This makes 4IR an essential feature in this sector in improving the scale of production and ensuring the commercialisation of farm products (Sung, 2018a).

Just like every other sector, 4IR is shaping the way farm products are being produced. Increased yields, reduced cost, and reduction in environmental impact are being experienced as a result of the introduction of digital technologies in farm activities. Furthermore, innovative ideas are being unravelled with new plant-based innovations and increasing resilience to extreme weather changes being evident in most countries' agricultural sector (Nijhuis & Herrmann, 2019). Field sensors connected to IoT are being employed to help farmers gather relevant data regarding soil nutrient and moisture leading to an informed decision that will help improve water usage and customise fertiliser blends. Also, the use of IoT helps eliminate the need for manual monitoring of greenhouses, since technology can help control the environment temperature, humidity, light levels, and at the same time conduct automatic irrigation. Big data and AI offer farmers a wide range of complex information needed for efficient decision-making. Scouting and monitoring of crops and livestock, inventory, precision spraying, and inspection of farm infrastructure are made easier through drones. Furthermore, monitoring crops for diseases by scanning them using visible and near-infrared lights has become easier through drones. These technologies offer the agricultural sector better productivity and profitability and in the creation of new locally based added value (Stankovic & Neftenov, 2020; Sung, 2018a).

Education

Within the education sector, the emergence of the concept "education 4.0" described as an approach of learning that is in sync with the values and features of 4IR (James, 2019) has become a popular paradigm. Education 4.0 is centred around the needs and potentials of 4IR in the education system, and learning models are designed to suit the real-time learners' profile (Almeida & Simoes, 2019; Popenici & Kerr, 2017). It has been stated that shortly, *"many of today's children will work in new job types that do not yet exist, with an increased premium on both digital and social-emotional skills in the coming years"* (World Economic Forum, 2020). To help prepare for this future challenging working environment, the use of technologies offered by 4IR is important. While the use of other 4IR technologies such as big data, sensors, and 3D printing might not be popular within the education sector, evidence of IoT, virtual reality (VR) and augmented reality (AR) has been mentioned within the sector (Shahroom & Hussin, 2018). Similarly, the use of serious games also known as applied games and gamification approaches in higher education projects has been noticed as some technologically advanced methods used in improving learning in higher education institutes (Almeida & Simoes, 2019). These technologies and approaches have recorded significant benefits over time.

The benefits of education 4.0 can be seen from two dimensions, namely the benefits of applying technology to improve learning and teaching and the benefit of the education system producing the needed workforce that will align with machines to explore new possibilities required in the ever-changing work environment. Furthermore, with education 4.0, teachers and students are benefitting from the adoption of technologies in learning. Teachers enjoy boarder knowledge of the course content they teach and at the same time are better informed as to how students assimilate what they teach through the use of AI-based portals. The students also now have more autonomy in learning (Almeida & Simoes, 2019; Demartini & Benussi, 2017).

Since 4IR comes with complex technologies, higher education and cognitive skills will be crucial for operations in the current industrial revolution. It is believed that countries around the world, especially the developing ones, will have to redefine their education focus and adopt models that will help them to be responsive if they are to enjoy economic growth, job creation, better safety and security, and improved knowledge transfer which the 4IR promises (Khatu, 2019). In the view of Xing and Marwala (2017), the higher education system in the era of 4IR can be complex, dialectic, and at the same time an exciting opportunity which can bring significant transformation to every society. But for this transformation to occur, interdisciplinary teaching, research and innovation have become more important than ever. Adopting wearables-assisted teaching and learning, embracing massive open online courses, cultivating innovative talent, and generalising blended learning have all been suggested as solutions for better teaching in the era of 4IR. In terms of research, promoting open, evolutionary and revolutionary innovations, research and development centred around new technological advancement, and shortened innovation cycles are suggested (*ibid.*).

Construction

The concept of 4IR is no longer regarded as a future trend as companies have now started to include it as part of their research and strategic agenda. Moreover, organisations are now transforming their businesses through integrating intelligent algorithms, advanced connectivity and automation, IoT, sensors, 3D printing, computer-powered processes, cloud computing, and connected capability (PricewaterhouseCoopers, PwC, 2016). Though other industries are transforming their activities through implementing technological advances and enjoying better productivity, efficiency, accuracy, and better customer satisfaction, the story is different for the construction industry. Despite the call by researchers and professionals for the adoption of "construction 4.0", the pace of embrace of digital tools is slow with traditional principles still common in the industry (Aghimien *et al.*, 2019; Alaloul *et al.*, 2020; Oesterreich & Teuteberg, 2016; Osunsanmi *et al.*, 2018). Implementation issue has been noted as one of the primary reasons for the slow pace of digital tools in the construction industry. In developing countries, however, it is believed that lack of trained personnel to handle these technologies, high cost of upskilling existing personnel, high cost of acquiring these technologies, absence of digital culture in the industry, resistance to change as well as organisational culture are some of the major problems hindering the implementation of the 4IR concept (Aghimien *et al.*, 2019; Alaloul *et al.*, 2020; Oke *et al.*, 2018; PwC, 2016). Dallasega *et al.* (2018) assert that although 4IR originated from the manufacturing industry, this technological advancement would transform the construction industry if given a chance. However, the industry still faces different challenges compared to the manufacturing industry. For example, construction projects require high degree of customisation and have a time period for completion, and project locations can be affected by weather conditions. Moreover, the distinctive characteristics of the construction industry, such as projects being site-based as well as a large number of small- and medium- enterprises firms with lack of resources to invest in the new technologies contribute to the slow adoption of 4IR (Dallasega *et al.*, 2018; Osunsanmi *et al.*, 2018).

If the industry can overcome some of the obstacles it faces, significant benefits await in the use of 4IR technologies. The adoption of 4IR can help construction firms reduce uncertainty and complexity and improve further communication and information exchange between parties involved in the project, resulting in improved quality and productivity. Also, reduced labour costs, reduced material cost, time-saving, improved building quality and enhanced work safety, and improved image are some of the benefits that the industry stands to gain in the long run from 4IR adoption (Dallasega *et al.*, 2018; Oesterreich & Teuteberg, 2016). The construction industry and its adoption of 4IR principles, particularly digitalisation, is discussed further in Chapter 4.

Digitalisation as a driver for 4IR

Based on the earlier description of 4IR, it is evident that this industrial revolution is hinged on new, emerging, and disruptive technologies. It has been observed that 4IR is driven by physical, digital, and biological technologies (Li *et al.*, 2017; Schwab, 2017) without which the inherent benefits of the industrial revolution cannot be achieved. Although continuous development is being done in these three different areas, a breakthrough in these individual drivers and their ability to interconnect makes 4IR different from the previous industrial revolutions. Furthermore, while the third industrial revolution emanated from hardware, 4IR was born from the breakthrough in software. Notable fields within these drivers are:

(1) Digital – IoT, AI, machine learning, big data, cloud computing, and digital platforms
(2) Physical – autonomous cars and 3D printing
(3) Biological – genetic engineering and neurotechnology (Li *et al.*, 2017).

This book focuses on the digital and physical aspects of 4IR.

In manufacturing, from where the concept of 4IR emanated, IoT, Industrial IoT, cloud-based and smart manufacturing have become prominent drivers for successful production (Erol *et al.*, 2016). Creating a smart factory and judiciously using the internet are vital ways to employ 4IR in the manufacturing business (Bédard-Maltais, 2017). Erol *et al.* (2016) observed that the internet and supporting digital technologies serve as a mainstay for the proper integration of human and machine agents, materials, products, production lines, and processes to form more intelligent, connected, and agile value chain. These digital technologies help transform the manufacturing industry's processes and products into an entirely digitised and intelligent one. This, in essence, implies that the concept of 4IR relies on digitalisation for it to be attained. In confirmation, Bienhaus and Haddud (2018) affirmed that digitalisation is an important driver for 4IR. However, the approaches towards achieving digital transformation in organisations differ in how organisations picture their inherent opportunities and handle their challenges. Therefore, the role of digitalisation in the attainment of 4IR cannot be overemphasised as it is the wheel that drives the industrial revolution.

It is important to note that while most literature have used the word "digitisation", "digital technologies", and "digitalisation" interchangeably, there is some level of distinction between these words.

Digitisation

Digitisation is a process that symbolically converts analogue signals into bits that are represented with zeros and ones. It gives information capable of being expressed in different ways, on different materials, and in different systems (Brennen & Kreiss, 2014). All forms of data such as alphanumeric text, graphics, still and moving pictures, and sounds can be digitised (Verhulst, 2002). Simply put, digitisation is the conversion of all analogue-related issues within an organisation, ranging from customer files, repair handbooks, ledgers, and the like into a digital format that can be easily accessed and transported through a digital device (Irniger, 2017).

Digital technologies

Digital technologies are the backbone of 4IR as nearly every innovation and advance in the industrial revolution is enhanced through digital power (Schwab, 2016). These technologies are electronic tools, systems, devices, or resources capable of creating, storing, and processing data. They depend on microprocessors and function through binary computational codes. They include computerised devices and applications that are dependent on computers such as mobile phones, video cameras, tablets, laptops, computers, internet, and communication satellites (Anjos-Santos *et al.*, 2016; Pullen, 2009). Therefore, Dimick (2014) concluded that digital technologies could be viewed from three facets, and these are software, information technology (IT) equipment (computers and related hardware), and communications equipment.

Digitalisation

Digitalisation is less concerned with zeros and ones and more concerned with how these zeros and ones shape people's social and economic lives. Crampton (2016) defined digitalisation as "*the use of digital technology to change the business model, provide new revenue streams and value-producing opportunities*". Digitalisation is the increased use of digital technologies in an organisation's operations (Kalavendi, 2017). It is believed to be the most important technological advancement, the impact of which will be felt at all levels of society (Leviakangas *et al.*, 2017). Dall'Omo (2017) observed that digitalisation had found an abode in every aspect of human life, be it simple personal devices or complex industrial systems. Ochs and Riemann (2018) noted that digitalisation incorporates digital technologies into everyday life, through digitising anything capable of being digitised. According to Leviakangas *et al.* (2017), a unique technological base for metasystems of products is created using intelligent sensors, robotics, and automation through digitalisation. Therefore, while digitisation is seen as the digital transformation of specific processes that were not digital, digitalisation is the deployment of these transformed processes into the organisation's business life using digital technologies. This points to the fact that by going through digitalisation, an organisation will have to attain digital transformation by abandoning the old analogue ways (paper and processes) for a more digital-oriented approach.

Summary

In summary, the fourth industrial revolution is an advancement of the third with more software development to digitise existing, new, and emerging technologies. The technologies of 4IR are disrupting activities in every sector of the economy and construction is no exception. Although the adoption of these technologies is slow within the construction industry, their usage offers a solution to age-long problems of the industry. Digital technologies drive the 4IR, and the application of these technologies in organisational activities is known as digitalisation. If construction organisations can adopt the concept of digitalisation, they stand the chance to transform their processes and products into an entirely digitised and intelligent one as experienced by their counterparts in the manufacturing industry.

References

Aghimien, D.O., Aigbavboa, C.O. and Matabane, K. (2019). Impediments of the fourth industrial revolution in the South African construction industry. *Construction in the 21st Century, 11th International Conference (CITC-11)*, London, UK, September 9–11, pp. 318–324

Alaloul, W.S., Liew, M.S., Zawawi, N.A.W.A. and Kennedy, I.B. (2020). Industrial revolution 4.0 in the construction industry: Challenges and opportunities for stakeholders. *Ain Shams Engineering Journal*, 11(1): 225–230

Almeida, F. and Simoes, J. (2019). The role of serious games, gamification and industry 4.0 tools in the education 4.0 paradigm. *Contemporary Educational Technology*, 10(2): 120–136

Anjos-Santos, L.M., El Kadri, M.S., Gamero, R. and Gimenez, T. (2016). Developing English language teachers' professional capacities through digital and media literacies: A Brazilian perspective. In *Manzoor, a. Editions of Handbook of Research on Media Literacy in the Digital Age*. IGI Publishers, Harrisburg, PA

Bahrin, M.A.K., Othman, M.F., Nor, N.H. and Azli, M.F.T. (2016). Industry 4.0: A review on industrial automation and robotic. *Jurnal Teknologi (Sciences and Engineering)*: 137–143

Bédard-Maltais, P. (2017). Industry 4.0: The new industrial revolution – are Canadian manufacturers ready? *BDC Study*. Available at: www.bdc.ca/EN/Documents/analysis . . . / bdc-resume-etude-manufacturing-en

Bienhaus, F. and Haddud, A. (2018). Procurement 4.0: Factors influencing the digitisation of procurement and supply chains. *Business Process Management Journal*, 24(4): 965–984

Brennen, S. and Kreiss, D. (2014). Digitalisation and digitization. *Culture Digitally*. Available at: http://culturedigitally.org/2014/09/digitalization-and-digitization/ [accessed 21–11–2020]

Celaschi, F. (2017). Advanced design-driven approaches for an industry 4.0 framework: The human-centred dimension of the digital industrial revolution. *Strategic Design Research Journal*, 10(2): 97–104

The Centre of Excellence in Financial Services, COEFS. (2017). *The Impact of the 4th Industrial Revolution on the South African Financial Services Market*. Centre of Excellence in Financial Services Report. Available at: www.genesis-analytics.com/uploads/down loads/COEFS-TheimpactofthefourthindustrialrevolutiononfinancialservicesinSouthAf rica-final-1-FR.pdf

Crampton, N. (2016). Swot analysis of the construction sector in South Africa. Available at: https://bizconnect.standardbank.co.za/sector-news/construction/swot-analysis-of-the-construction-sector-in-south-africa.aspx [accessed 15–8–2018]

Crnjac, M., Veža, I. and Banduka, N. (2017). From concept to the introduction of industry 4.0. *International Journal of Industrial Engineering and Management*, 8(1): 21–30

Dallasega, P., Rauch, E. and Linder, C. (2018). Industry 4.0 as an enabler of proximity for construction supply chains: A systematic literature review. *Computers in Industry*, 99: 205–225

Dall'Omo, S. (2017). Driving African development through smarter technology. *African Digitalisation Maturity Report*: 1–45

Demartini, C. and Benussi, L. (2017). Do Web 4.0 and industry 4.0 imply education X.0? *IT Pro*, 4–7 May/June.

Dimick, S. (2014). *Adopting Digital Technologies: The Path for SMEs*. The Conference Board of Canada, Ottawa, ON, pp. 1–13

Erol, S., Jäger, A., Hold, P., Ott, K. and Sihn, W. (2016). Tangible industry 4.0: A scenario-based approach to learning for the future of production, 6th CLF – 6th CIRP Conference on learning factories. *Procedia CIRP*, 54: 13–18

Hermann, M., Pentek, T. and Otto, B. (2016). Design principles for industries 4.0 scenarios. *49th Hawaii International Conference on System Sciences (HICSS)*, Koloa, HI, pp. 3928–3937

Irniger, A. (2017). Difference between digitization, digitalization and digital transformation. Available at: www.coresystems.net/blog/difference-between-digitization-digitalization-and-digital-transformation

James, F. (2019). Everything you need to know about education 4.0. Available at: www.qs.com/everything-you-need-to-know-education-40/ [accessed 14–12–2020]

Kagermann, H. (2014). *How Industry 4.0 Will Coin the Economy of the Future? The Results of the German High-Tech Strategy's and Strategic Initiative Industrie 4.0*. Royal Academy of Engineering, London

Kalavendi, R. (2017). Digitisation and digitalisation: The two-letter difference. Available at: https://medium.com/@ravisierraatlantic/digitization-digitalization-the-2-letter-dif ference-59b747d42ade [accessed 19–11–2020]

Khatu, R. (2019). 4IR and the South African education system. Available at: www.bizcom munity.com/Article/196/371/193213.html#:~:text=The%20fourth%20is%20a%20 fusion,education%20levels%20and%20cognitive%20skills [accessed 19–11–2020]

Klima, T. (2019). The fourth industrial revolution for finance. Available at: www.digital istmag.com/finance/2019/01/31/fourth-industrial-revolution-for-finance-06194670/ [accessed 4–11–2020]

Lee, J. (2013). Industry 4.0 in big data environment. *Germany Harting Magazine*, 1: 8–10

Leviakangas, P., Paik, S. and Moon, S. (2017). Keeping up with the pace of digitisation: The case of the Australian construction industry. *Technology in Society*, 50: 33–43

Li, G., Hou, Y. and Wu, A. (2017). Fourth industrial revolution: Technological drivers, impacts and coping methods. *Chinese Geographical Science*, 27(4): 626–637

Nair, V. (2019). It's time for financial services to embrace the fourth industrial revolution. *Here's Why*. Available at: www.weforum.org/agenda/2019/03/its-time-for-financial-services-to-embrace-the-fourth-industrial-revolution-heres-why/ [accessed 4–11–2020]

Nijhuis, S. and Herrmann, I. (2019). The fourth industrial revolution in agriculture. *Strategy Business*. Available at: www.strategy-business.com/article/The-fourth-industrial-revolution-in-agriculture?gko=75733 [accessed 10–12–2020]

Ochs, T. and Riemann, U.T. (2018). IT strategy follows digitalisation. In *Encyclopedia of Information Science and Technology*, 4th edition. IGI Publishers, PA

Oesterreich, T.D. and Teuteberg, F. (2016). Understanding the Implications of Digitisation and Automation in the Context of Industry 4.0: A Triangulation Approach and Elements of a Research Agenda for the Construction Industry. *Computers in Industry*, 83: 121–139

Ohene-Afoakwa, E. and Nyanhongo, S. (2017). *Banking in Africa: Strategies and Systems for the Banking Industry to Win in the Fourth Industrial Revolution*. African Expansion Project. Available at: www.bankseta.org.za/wp-content/uploads/2018/08/BA3DD51-1.pdf [accessed 19–11–2020]

Oke, A.E., Aghimien, D.O., Aigbavboa, C.O. and Koloko, N. (2018). Challenges of digital collaboration in the South African construction industry. *Proceedings of the International Conference on Industrial Engineering and Operations Management*, Bandung, Indonesia, 6–8 March, pp. 2472–2482

Osunsanmi, T.O., Aigbavboa, C.O. and Oke, A.E. (2018). Construction 4.0: The future of South Africa construction industry. *World Academy of Science, Engineering and Technology International Journal of Civil and Environmental Engineering*, 12(3): 206–212

Petrillo, A., De Felice, F., Cioffi, R. and Zomparelli, F. (2018). Fourth industrial revolution: Current practices, challenges, and opportunities. Digital transformation in smart manufacturing digital transformation in smart manufacturing. In Petrillo, A., Cioffi, R. and Fabio De, F. (Eds.). IntechOpen. Available at: www.intechopen.com/books/digital-transformation-in-smart-manufacturing/fourth-industrial-revolution-current-practices-challenges-and-opportunities [accessed 19–11–2020]

Popenici, S. and Kerr, S. (2017). Exploring the impact of artificial intelligence on teaching and learning in higher education. *Research and Practice in Technology Enhanced Learning*, 12(22): 1–13

Pullen, D.L. (2009). Back to basics: Electronic collaboration in education sector. In Salmons, J. and Wilson, L. (Eds.), *Editions Handbook of Research on Electronic Collaboration and Organizational Synergy*. IGI Publishers, Harrisburg, PA

PwC. (2016). Clarity from above. PwC global report on the commercial applications of drone technology. Available at: https://pwc.blogs.com/files/clarity-from-above-pwc.pdf [accessed 14–12–2020]

Roblek, V., Mesko, M. and Krapez, A. (2016). A complex view of industry 4.0. *SAGE Open*, April–June: 1–11

Rüßmann, M., Lorenz, M., Gerbert, P. and Waldner, M. (2015). *Industry 4.0: The Future of Productivity and Growth in Manufacturing Industries*. Boston Consulting Group Report, Boston

Schwab, K. (2016). *The Fourth Industrial Revolution*. World Economic Forum, Switzerland, pp. 25–38

Schwab, K. (2017). *The Fourth Industrial Revolution*, 1st edition. Crown Business, New York

Shahroom, A. and Hussin, N. (2018). Industrial revolution 4.0 and education. *International Journal of Academic Research in Business and Social Sciences*, 8(9): 314–319

Stankovic, M. and Neftenov, N. (2020). The fourth industrial revolution and its potential applications in agriculture in Africa. *Agrilinks*. Available at: www.agrilinks.org/post/fourth-industrial-revolution-and-its-potential-applications-agriculture-africa [accessed 10–12–2020]

Stefanul, A. (2020). Industry 4.0 and its impact on the financial services. *FinTech Weekly*. Available at: https://fintechweekly.com/magazine/articles/industry-4-0-and-its-impact-on-the-financial-services [accessed 4–11–2020]

Sung, J. (2018a). The fourth industrial revolution and precision agriculture. In Hussmann, S. (Ed.), *Automation in Agriculture*. Intech, Chapter 1. http:// dx.doi.org/10.5772/intechopen.71582

Sung, T.K. (2018b). Industry 4.0: A Korea perspective. *Technological Forecasting and Social Change*, 132(C): 40–45

Tanda, A. and Schena, C. (2019). *FinTech, BigTech and Banks – Digitalisation and Its Impact on Banking Business Models*. Palgrave Macmillan Studies in Banking and Financial Institutions, Springer Nature AG, Chem, Switzerland

Vaidya, S., Ambad, P. and Bhosle, S. (2018). Industry 4.0 A glimpse. *Procedia Manufacturing*, 20(11–12): 233–238

Verhulst, S. (2002). About scarcities and intermediaries: The regulatory paradigm shift of digital content reviewed. In Lievrouw, L.A. and Livingstone, S. (Eds.), *The Handbook of New Media*. Sage, London, pp. 432–447

World Economic Forum. (2020). Education 4.0. Available at: www.weforum.org/projects/learning-4-0 [accessed 14–12–2020]

Xing, B. and Marwala, T. (2017). Implications of the fourth industrial age on higher education. *Thinker: For the Thought Leaders*, 73: 10–15

4   Digitalisation in construction

Introduction

This chapter explores the concept of construction digitalisation by combing through past literature to unearth their area of focus and determine how digitalisation has been viewed in the construction context. The chapter adopted a scientometric review of existing studies to achieve this feat. Based on the scientometric study and definition of digitalisation, the chapter defines the construction digitalisation and discusses the different digital technologies that construction organisations can adopt in their quest for digital transformation. The chapter also explores some of the drivers, barriers, impacts, and risks of construction digitalisation.

Understanding construction digitalisation through past research

To understand the views of digitalisation in construction, a scientometric review of existing studies was conducted. Key areas of concern in published works on digitalisation within construction-related fields were assessed. Due to the newness of the concept of digitalisation in construction vis-à-vis its slow adoption, a scientometric review was deemed necessary. This will allow a concise identification and mapping of scientific knowledge areas by identifying research patterns and boundaries. Scientometrics (as a sub-branch of bibliometrics) is described as the quantitative approach used in text mining of scientific publications (Hawkins, 2001). A scientometric review allows the identification of popular authors, country of origin, funding bodies, journals, and types of collaborations within a community of practice that studies a common phenomenon (Blažun et al., 2015). To achieve this, the Scopus database was used to search pertinent literature – a decision premised upon the fact that Scopus covers a considerably larger number of published articles than its counterparts (Aghimien et al., 2019; Guz & Rushchitsky, 2009). Using one accessible database also helps reduce the problem of duplication of the extracted article.

Using the Scopus database, the search for relevant literature focused on published journal articles and conference proceedings within the study area of engineering (which includes construction). The choice of selecting journal articles was because journals provide the most scientifically reliable sources of knowledge having been subjected to a rigorous review process (Jin *et al.*, 2018; Zheng *et al.*, 2016). However, while some researchers advocate the use of only journal articles for literature review (Butler & Visser, 2006; Jin *et al.*, 2018; Zhao, 2017), others have favoured the inclusion of conference proceedings particularly in rapidly evolving areas such as information and digital sciences (Vuksic *et al.*, 2018; Webster & Watson, 2002). The search for relevant literature was undertaken using TITLE-ABS-KEY – 'digitalisation' OR 'digitisation' OR 'digital technologies' OR 'digital transformation', AND 'construction'. The search timeline was ten years between 2010 and 2020. This is because the concept of 4IR started gaining popularity only since the 2011 Hannover Fair in Germany (Sung, 2018). Searching for documents from before this time might give little insights into the focus on digitalisation and its adoption in construction. Furthermore, articles published in only the English language were considered. The literature search was conducted in December 2020, and the initial search revealed a total of 5,314 published documents. After careful refining of the articles based on the study area, language, and publication type, a total of 863 papers were deemed fit for further assessment.

Visualisation of Similarities viewer (VOSviewer) was used to develop the co-occurrence network of keywords. While several scientometric analysis software tools were considered (e.g., BibExcel, CiteSpace, Gephi), VOSviewer was adopted because it is comparatively user-friendly, free to access and has an inherent ability to generate intuitive results (Nazir *et al.*, 2020). Moreover, it has been extensively used in prevailing architecture, engineering and construction (AEC) literature (Aghimien, Aigbavboa, Oke, & Thwala, 2019; Jin *et al.*, 2019; Darko *et al.*, 2020). The software also enables the easy visualisation of bibliometric networks by displaying nodes and providing distance-based visualisations (van Eck & Waltman, 2014). Furthermore, Shrivastava and Mahajan (2016) state that using a network of keywords offers a better pictorial view of a knowledge area and provides a clearer understanding of research interest.

Publication per year and types

Out of the 860, extracted documents conference proceedings had 550 documents, while 313 were articles in reputable journals. This result can be attributed to the rigour and long duration taken in the publication process of journal articles compared to conference papers. Hence, with longer publication and review process come fewer publications within journals. Figure 4.1 reveals a gradual increase from 53 documents in 2010 to 146 in the year 2014. This is expected as 4IR which is driven by digitalisation started gaining significant recognition only since 2011, and issues surrounding 4IR became topical in most industries due to its focus on creating a smart environment (Crnjac *et al.*, 2017). However, there was a drastic drop in the number of publications in 2015 as the year recorded only 45 documents. The year 2016 through 2019 experienced a steady rise in publications with 2019

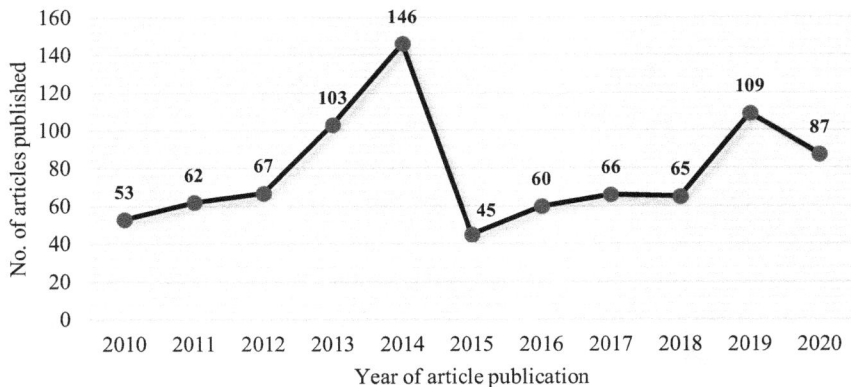

Figure 4.1 Number of digitalisation-related articles per publication year

recording 109 papers. However, the year 2020 recorded a drop in the number of publications as only 87 documents were extracted.

Top countries contributing to digitalisation in construction research

The extracted documents emanated from 61 countries. Some of these countries have only one document that has not gained any citation; hence their contribution to the construction digitalisation discourse cannot be ascertained. To get a clearer picture of those countries promoting the digitalisation of construction activities, the minimum number of documents per country was set at five. The number of citations was set at ten, and the result is presented in Figure 4.2. The figure shows that China tops the list with a total of 414 research publications and 813 citations. This is followed by the USA (107 publications and 1227 citations), Germany (49 publications and 6147 citations), and the UK (45 publications and 503 citations). Australia, Canada, and Japan also had at least 20 documents with significant number of citations. Furthermore, a cursory look at the figure shows that top countries contributing to digitalisation in construction research are within Asia, America, and Europe. This submission is in tandem with Olawumi *et al.* (2017), who observed that most research on digital technologies such as BIM (a crucial construction digitalisation feature) published between 1990 and 2016 emanated from Asia and Europe.

Similarly, Lee and Yu (2016) noted that the Korean central government strongly encourages construction digitalisation using BIM. This is supported by the creation and enforcement of BIM policies and funding of BIM research. A similar observation was noted in the study of Won *et al.* (2013), which noted a significant increase in the adoption of digitalisation related tools in the USA and Korea. Vuksic *et al.* (2018) also submitted that digital transformation

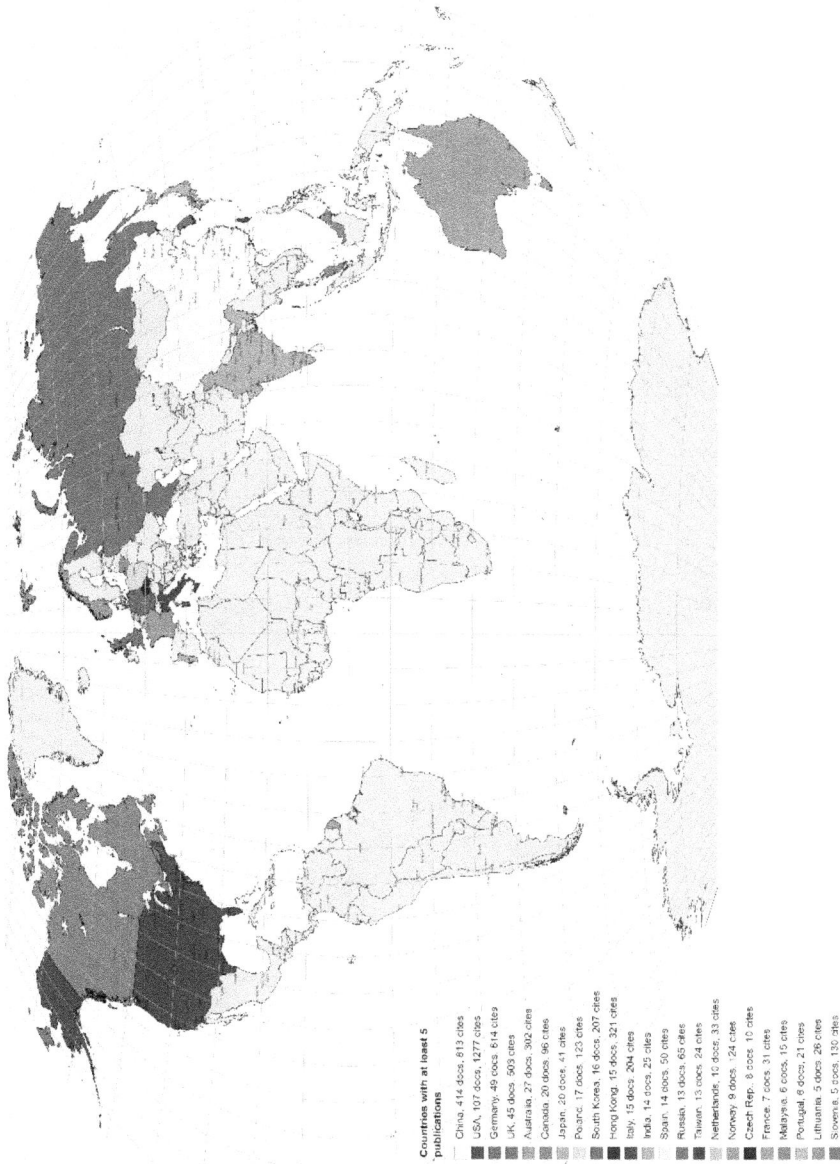

Countries with at least 5 publications

China, 414 docs, 813 cites
USA, 107 docs, 1277 cites
Germany, 49 docs, 614 cites
UK, 45 docs, 503 cites
Australia, 27 docs, 302 cites
Canada, 20 docs, 96 cites
Japan, 20 docs, 41 cites
Poland, 17 docs, 123 cites
South Korea, 16 docs, 207 cites
Hong Kong, 15 docs, 321 cites
Italy, 15 docs, 204 cites
India, 14 docs, 25 cites
Spain, 14 docs, 50 cites
Russia, 13 docs, 65 cites
Taiwan, 13 docs, 24 cites
Netherlands, 10 docs, 33 cites
Norway, 9 docs, 124 cites
Czech Rep, 8 docs, 10 cites
France, 7 docs, 31 cites
Malaysia, 6 docs, 15 cites
Portugal, 6 docs, 21 cites
Lithuania, 5 docs, 26 cites
Slovenia, 5 docs, 130 cites

Figure 4.2 Countries with at least two publications on digitalisation

Source: Author's compilation using mapchart.net

research is more from European countries (such as Germany and Baltic), the USA, and Canada. The Institute of Management Development (IMD) "world digital competitiveness" ranked the USA as the most digitally competitive country in the world (IMD, 2018). This is evident in the high numbers of research emanating from the country and the high number of citations recorded in this study.

Interestingly, based on the threshold of a minimum of five documents and ten citations, no country in Africa can be found on the map (refer to Figure 4.2). Further investigation revealed that Nigeria has two papers that have been cited twice. A journal publication also emanated from Algeria but has never been cited, while Tunisia recorded one conference paper that has been cited 11 times. South Africa also recorded one publication cited twice as a result of a collaboration done in one of the publications from Nigeria. This dearth of research conducted in Africa reveals the potential for AEC researchers to conduct scientific work in this area for the socio-economic benefit of developing countries.

Most cited publications

In recent times, citation counts have been used to determine the bibliometric performance of scholarly publications and academics' impact within their respective academic community (Hirsch, 2005; Wang *et al.*, 2019; Zhang, 2013). Table 4.1

Table 4.1 Most cited publications

Document	Title	Source	Cites	Focus	Approach
Siebert and Teizer (2014)	Mobile 3D mapping for surveying earthwork projects using an Unmanned Aerial Vehicle (UAV) system	*Automation in Construction*	393	Performance of digital technology (UAV)	Experiment
Pauwels and Terkaj (2016)	EXPRESS to OWL for construction industry: Towards a recommendable and usable ifcOWL ontology	*Automation in Construction*	144	Information management	Experiment
Heydarian *et al.* (2015)	Immersive virtual environments versus physical built environments: A benchmarking study for building design and user-built environment explorations	*Automation in Construction*	116	DTs for design, construction and operation of building projects	Experiment

Document	Title	Source	Cites	Focus	Approach
Zhou *et al.* (2012)	Construction safety and digital design: A review	*Automation in Construction*	114	Construction safety	Review
Bhatla *et al.* (2012)	Evaluation of accuracy of as-built 3D modelling from photos taken by handheld digital cameras	*Automation in Construction*	103	3D modelling	Experiment
Kim *et al.* (2015)	A framework for dimensional and surface quality assessment of precast concrete elements using BIM and 3D laser scanning	*Automation in Construction*	93	DTs for buildings structures (BIM and laser scanner)	Case study
Smith (2014)	BIM implementation –global strategies	*Procedia Engineering*	86	BIM adoption	Review
Irizarry *et al.* (2013)	InfoSPOT: A mobile augmented reality method for accessing building information through a situation awareness approach	*Automation in Construction*	82	DTs for facility management (AR)	Experiment
Turk and Klinc (2017)	Potentials of blockchain technology for construction management	*Procedia Engineering*	76	DTs for construction Information Management (Blockchain)	Review
Li, Xue *et al.* (2018)	An Internet of Things-enabled BIM platform for on-site assembly services in prefabricated construction	*Automation in Construction*	73	DTs for prefabricated construction (IoT and BIM)	Experiment
Zhu *et al.* (2010)	Large-scale geomechanical model testing of an underground cavern group in a true three-dimensional (3-D) stress state	*Canadian Geotechnical Journal*	61	Underground construction	Experiment
Delgado Camacho *et al.* (2018)	Applications of additive manufacturing in the construction industry – A forward-looking review	*Automation in Construction*	60	3D Printing adoption	Review

(*Continued*)

Table 4.1 (Continued)

Document	Title	Source	Cites	Focus	Approach
Dainty *et al.* (2017)	BIM and the small construction firm: a critical perspective	Building Research and Information	53	BIM adoption	Review
Meža *et al.* (2014)	Component based engineering of a mobile BIM-based augmented reality system	*Automation in Construction*	53	BIM and AR for construction delivery	Experiment
Woodhead *et al.* (2018)	Digital construction: From point solutions to IoT ecosystem	*Automation in Construction*	51	Evolution of DTs in construction	Review

Source: Author's compilation (2020)

reveals 15 publications with at least 50 citations. From the table, the most cited publication with 393 citations is the work of Siebert and Teizer (2014), which adopted an experimental approach in surveying earthwork projects using Unmanned Aerial Vehicle (UAV) system. The article was published in automation in a construction journal rated Q1 in the "engineering" domain with an H-index of 107. The journal publishes original articles focusing on diverse aspects relating to the use of IT in design, engineering, construction technologies, and maintenance and management of constructed facilities (Scimago Journal Rank, 2020). Therefore, it is not surprising to see 11 out of the 15 documents on the table emanating from this journal. Pauwels and Terkaj's (2016) study places emphasis on information management and exchange in the construction industry. The paper has garnered 144 citations and was also published in the automation in construction journal. This is followed by the works of Heydarian *et al.* (2015) and Zhou *et al.* (2012) with 116 and 114 citations respectively. While the former focused on immersive virtual and physical built environments in creating benchmarks for building design and user-built environment explorations, the latter explored construction safety through digital design. Evidently, the tables show that most cited studies have explored the performance of digital technologies, the general adoption of these technologies, and their application in solving specific identified problems within the construction industry. Furthermore, experimental approaches have been favoured as an ideal methodology in most studies. This is a pointer to the fact that future research can still be done in some of these areas using other methods or a combination of methods.

Main research interest on digitalisation in construction

Past studies' focus was analysed using the co-occurrence of keywords from the scientometric data extracted using VOSviewer. This study considered all keywords (i.e., authors' keywords and the journals' indexed keywords) for network visualisation to ensure full and comprehensive coverage. The assessed documents produced a total of 8,136 keywords. VOSviewer helps to group these keywords into clusters using a set minimum threshold for co-occurrences. These clusters clarify concentration areas of investigation in past studies. The software has a default minimum number of co-occurrence for the keywords to be extracted as five; hence, a keyword must be mentioned in at least five different papers to appear on the map (Nazir *et al.*, 2020). However, several previous studies have adopted different thresholds based on the number of documents under review. For example, Darko *et al.* (2020) used 50 for 41,827 extracted documents on artificial intelligence in the AEC industry. Yin *et al.* (2019) adopted 20 co-occurrences for 4,395 extracted documents on BIM for off-site construction. Aghimien, Aigbavboa, Oke, and Thwala (2019) adopted four as the minimum for reviewing 91 articles on robotics and automation in construction-related studies. Saka and Chan (2019) used two as the minimum number of co-occurrence for their review of 93 documents on BIM-related studies in Africa. Cumulatively, these studies illustrate a lack of unified consensus regarding the ideal number of minimum co-occurrence to be adopted. Therefore, this current study adopted ten co-occurrences for a keyword to be extracted due to many keywords in the assessed documents.

Furthermore, this minimum threshold gave an optimal graphic presentation of the network maps. A total of 136 keywords met this threshold with a total link strength (TLS) of 3,321. Figure 4.3 illustrates the network visualisation map with five clusters representing each area of focus of digitalisation in construction research.

Cluster 1 – arbitrarily titled "**data and information management**" is evident within the region marked as cluster 1 on the map and has 37 extracted keywords. Prominent among these keywords are digital storage, information management, information technology, digital libraries, digital campus, data acquisition, information services, computer technology, big data, cloud computing, information science, manufacture, and design.

Cluster 2 – arbitrarily titled "**digital technologies in building design, construction and operation**" is evident within the region marked as cluster 2 on the map and has 30 extracted keywords. Among these keywords are architectural design, construction industry, project management, digital technologies, building information model – BIM, construction, construction, life cycle, and information theory.

Cluster 3 – arbitrarily titled "**digital technologies in civil and heavy engineering**" is evident within the region marked as cluster 3 on the map and has 29 extracted keywords. Among these keywords are structural design, computer-aided design, concrete construction, concretes, product design, 3D printing, image processing, digital fabrication, reinforced concrete, computer simulation, optimisation, concrete buildings, bridges, tunnels, safety engineering, machinery, among others.

Cluster 4 – arbitrarily titled "**digital technologies for sustainable and smart construction**" is evident within the region marked as cluster 4 on the map and has 27 extracted keywords. Among these keywords are construction equipment, virtual reality, 3D computer graphics, IoTs, automation, visualisation, photogrammetry,

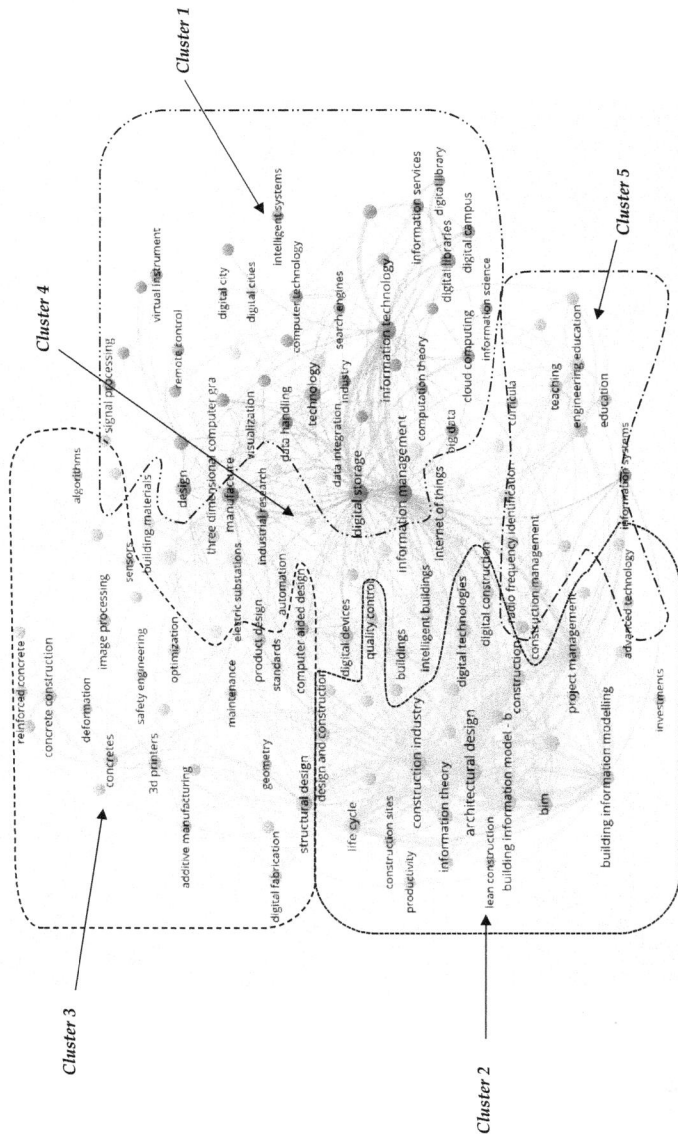

Figure 4.3 Co-occurrence network map

Source: Author's compilation using VOSviewer

intelligent buildings, standards, digital cities, network architecture, data integration, data security, sustainable development, standards, urban planning among others.

Cluster 5 – arbitrarily titled **"Skills development for digital technology usage"** is evident within the region marked as cluster 5 on the map and has 13 extracted keywords. These keywords are engineering education, students, digital devices, construction management, teaching, education, AI, e-learning, human resource management, curricula, learning systems, advance technology, and construction engineering.

Based on the preceding, it is evident that significant attention has been given to the management of information through digital technologies for retrieving, processing, and storing gathered data. Furthermore, the emphasis is on how digital technologies can be used right from the design phase through to the operation of construction projects. This is more of a generic view of the diverse technologies needed in the delivery of construction products and has become necessary as the construction industry has been noted for its slow adoption of digital technologies. The application of digital technologies in civil and heavy engineering projects has also been explored. This is crucial as machines are more capable of delivering more dangerous projects instead of using manual labour. Therefore, exploring how these technologies can help better deliver dangerous and complex projects is necessary. In the same view, some studies have beamed their searchlight on digital technologies that help provide more sustainable projects within the industry and provide smart and intelligent projects to construction clients in the process. Finally, since technologies need to work side by side with man, other studies have focused on how existing skills can be improved to have the required skills that can help attain the full potential of digital technologies in the construction industry.

It is pertinent to note that the result of scientometric review gives insight only into the research area of focus in construction digitalisation. Its submission should not be generalised as the only areas of focus as only one database was explored. Instead, the result should be used to understand the major areas researchers have explored and the gap in existing construction digitalisation knowledge like the absence of studies on the maturity of the construction industry and its organisations.

Defining construction digitalisation

Earlier in Chapter 3, the description of digitalisation was given as the use of digital technologies in organisations' daily activities by digitising anything capable of being digitised. This use of digital technologies allows the transformation of organisations' business model while allowing for new revenue streams to be derived and creating value-producing activities (Crampton, 2016; Kalavendi, 2017). In confirmation, the scientometric review shows that there will be no digitalisation without the proper adoption of digital technologies as all clusters of keywords point to the use of one digital technology or another in achieving diverse construction activities. Looking at the earlier description of digitalisation vis-à-vis research areas unearthed from the scientometric review conducted, it is necessary to define construction digitalisation in the context of this study. The study draws from Aghimien, Aigbavboa, and Oke's (2019) definition of construction digitalisation to achieve this. Thus, construction digitalisation in the context of this book is *"the innovative use of digital technologies in the delivery of tangible and intangible services within a construction organisation to gain a competitive advantage over other competitors while providing better service delivery"*.

Digital technologies needed for construction digitalisation

Understanding the digital technologies applicable to construction digitalisation and how they have been adopted over the years is necessary to clarify the context of digitalisation in the industry.

Building information modelling

Building information modelling (BIM) is drastically changing planning, designing, and managing buildings, infrastructures, and utilities. BIM has been described as a platform that promotes collaboration between project stakeholders and at the same time allows for a significant improvement in project delivery (Abubakar et al., 2014). Wong and Fan (2013) have earlier stated that BIM is a new and innovative technology, which has made the attainment of sustainable designs more efficient in recent times. However, Gilkinson et al. (2015) argued that while BIM can be seen as a new and innovative technology within the AEC industry, the concept has been in existence for well over 30 years. Eastman et al. (2011) also noted that what is today known as BIM can be dated back to the 1970s, but the terminology "Building Information Modelling" has been in existence for at least 15 years. Gilkinson et al. (2015) further stated that BIM has existed under different names such as building product models. Its usage is evident in the aerospace and automotive industry in analysing their products' design, performance, and manufacturability. These models have given tremendous result to these industries over time.

BIM has gained significant attention within the AEC industry due to its ability to provide easy use and reuse of project data across the project development phases. Simultaneously, it prevents unnecessary replication of project or design tasks (Kovačić et al., 2015; Lee & Yu, 2016; Olawumi et al., 2017). The Associated General Contractors of America (2006) defined BIM as *"the development and use of computer software model to simulate the construction and the operation of a facility"*. Eastman et al. (2011) submitted that BIM is *"a new approach to design, construction, and facilities management, in which a digital representation of the building process is used to facilitate the exchange and interoperability of information in digital format"*. Mahalingam et al. (2015) also described BIM as an innovative digital platform that aids information sharing among project participants and improves project performance in the process. Therefore, it can be said that BIM is a digital innovation designed to attain easy sharing of project information among project team members all through the life cycle of a project to attain better project performance through informed and reliable decision-making.

BIM has found its way into leading AEC organisations in most developed countries but at a significant slow pace in developing ones (Bui *et al.*, 2016; Jayasena & Weddikkara, 2013; Saka *et al.*, 2020). Haron *et al.* (2010) observed that most organisations within the built environments worldwide are moving towards adopting BIM in their practices. In the UK, large-sector stakeholders have changed their working processes, roles, and methods to accommodate BIM practices (Gilkinson *et al.*, 2015). Even the government has been a vocal supporter of establishing BIM in the construction industry (Howard *et al.*, 2017). Over time researchers and establishments have aspired to quantify BIM appropriation status since this is basic in assessing BIM adoption and other issues surrounding the technological innovation. This understanding has prompted reports such as the SmartMarket arrangement, National Building Specifications reports, BIM overviews, and others. These reports mostly studied respondents' involvement with BIM, advantages and future significance of BIM, possible utilisation of BIM programming applications, and BIM speculation (Succar, 2009). Based on the outcome of some of these reports, the selection rate and experience level of respondents in North America quickly expanded from 2009 to 2012, and the respondents from South Korea and Oceania also anticipated that their profundity of BIM execution would increase sooner rather than later (Jung & Lee, 2015).

In terms of its construction benefits, BIM provides a digitally constructed virtual model that can support the designing, procurement, fabrication, and construction activities needed for a project's realisation (Abubakar *et al.*, 2014). This virtual model could assist in facility maintenance and management during the operation phase of such a project (Azhar, 2011). Also, BIM is beneficial to almost all project participants involved in the different life cycles of a project. This is because it can create models that are useful throughout the project life cycle and simultaneously allow the parametric interaction (parametric modelling) of objects and components within the model. BIM also allows the easy identification of collisions and clashes in design, thereby allowing a reduction or total elimination of time and cost wastage that would have emanated from the construction of such faulty design. Also, it allows for easy estimation of cost and schedule of different elements within the model (Aboushady & Elbarkouky, 2015). With BIM, quantity surveyors' challenges in areas of cost estimating and quantity take-off are made easier. This is because BIM provides an avenue to automatically generate quantity take-offs and measurement directly from a digital model of a building (Cartlidge, 2013). Furthermore, BIM allows accurate shop drawings, easy design checks, and evaluation using software and provides an easy platform for sharing model information between all project participants (Aboushady & Elbarkouky, 2015). Besides, BIM offers increased productivity through effective communication and collaboration between all project stakeholders from inception to completion.

Although BIM poses numerous benefits to the construction industry, Aboushady and Elbarkouky (2015) noted that the invention is not without its problems. For example, there is no standard contract document for BIM in developing countries. There is also the need for optimum precision in the data being inputted to get positive results. Fear of modification of engineers' and architects' design by contractors and suppliers tends to affect the easy flow of information. According to Fox and Hietanen (2007), some of the challenges of adopting BIM are associated explicitly with BIM itself, while others are related to the general diffusion of innovation. Eastman *et al.* (2011) categorised these challenges into process challenges such as legal and organisational issues and technology challenges related to BIM readiness and implementation. This implies that most of the barriers to the adoption and effective use of BIM are process- and technology-related. Won *et al.* (2013) submitted that interoperability, investment, training, professional liability, process problems, trust, intellectual property, among others are affecting the worldwide implementation of BIM. For developing countries, the implementation of BIM in the construction industry is still slow. This problem has been attributed to lack of awareness, knowledge, and understanding of the BIM concept, lack of policy directives, and regulatory framework for its implementation, resistance to change, lack of government support, and high cost of implementation (Abubakar *et al.*, 2014; Addy *et al.* 2018; Jayasena & Weddikkara, 2013; Olugboyega & Aina, 2016; Saka & Chan, 2019; Yan & Damian, 2008).

Internet of Things

In the world today, the Internet has become an integral part of technological advancement and societal function. Advances made in ICT have led to a series of developments that are currently changing societies. One such improvement is IoT, which has been described as a global infrastructure designed for the information society to enable more advanced service delivery through the interconnectivity of physical and virtual objects using existing and frequently evolving interoperable ICTs (International Telecommunication Union, 2014). According to Hermann *et al.* (2016), IoT is the *"cyber-physical systems that communicate and cooperate and with humans in real-time via internet services, through which both internal and cross-organisational services are offered and used by participants throughout the value chain".* Dong *et al.* (2017) defined IoT as a *"network linking things that connects objects and the internet with a pre-set protocol using the information sensor equipment such as radio frequency identification technology (RFID), sensors and two-dimensional codes".* IoT makes communication between people and objects easier, faster, and more efficient (Magrassi & Berg, 2002). IoT has also been described as the interconnectivity between the internet and physical objects through a wireless means or via the use of cable networks plus unique identification codes with the sole purpose of mutually exchanging data (Atzori *et al.*, 2010; Gubbi *et al.*, 2013).

Porter and Heppelmann (2014) noted that IoT is special and different from the everyday internet because of the associated word "things" and what those things can do. The capabilities of IoT were grouped into four distinct areas, with one depending on the other. These capabilities are monitoring, control, optimisation, and autonomy. Using sensors and other external data sources, product condition, operations and usage, and external environment of the product are monitored. This gives room for the identification of areas of changes. In terms of control, there is software embedded in products to help control product function and at the same time personalise user's experience of such product. Product performance is enhanced through previous capabilities which enable algorithms that optimise product operations. This also allows predictive diagnostics, service, and repair. In addition, this previous capability ensures product operations are autonomous, allows self-coordination of operations with other products and systems, and creates autonomous product enhancement and personalisation.

IoT has grown over time since its introduction in developed countries, as it brings convenience in terms of living and consumption system of users, improves their quality of life, and reduces their workload (Dong *et al.*, 2017). Crnjac *et al.* (2017) noted that IoT is the backbone of 4IR as it helps connect people, things, and machines together. Stamford (2015) mentioned that by the year 2020, over 26 billion "things" will be connected to the internet. The development of IoT will, therefore, continue to evolve as research is being carried out on the technical aspects of design improvement (Gubbi *et al.*, 2013), application of IoT business models in organisations (Peoples *et al.*, 2013; Weber, 2010; Zhao *et al.*, 2013), IoT user's acceptance (Kim & Shin, 2015), and even social and environmental issues (Shin & Park, 2017). These research studies are geared towards improving IoT's capability in delivering a better way of life to people. However, Shin and Park (2017) observed that while IoT is rapidly developing globally, its associated problems cannot be overlooked. More significant is the issue of cyberattacks that come with the expanding interconnectivity of devices. This interconnectivity allows the linking of several aspects of human lives, and cyber attackers can gain access through security loopholes if not properly checked.

According to Ammar *et al.* (2018), the construction industry has embraced technologies such as mobile phones and computers connected via the internet. Urie (2019) further noted that IoT could be utilised in the construction industry to achieve self-diagnosis, self-configuration, and self-optimisation to enhance the service delivery process of the industry. Similarly, Ashton (2009) has stated that the construction industry can use IoT to track supplies, reduce waste, reduce losses and cost, and know when equipment needs to be repaired or replaced. Ammar *et al.* (2018) further recommended that the construction industry reinforces the IoT as the industry is likely to get more complex and complicated than it is currently. Just like BIM, the adoption and implementation of IoT within the construction industry of most developing countries are still slow and limited to very few applications. This implementation is impeded by issues bordering on lack of safety and security, lack of standards, lack of awareness of the inherent benefits to be derived, and poor connectivity (Gamil *et al.*, 2020).

Big data analytics

Effective digital transformation depends on several factors, among which is the electronic collection and analysis of digital data. Big data is a core driver of digital transformation in organisations (Zhong *et al.*, 2016). Bienhaus and Haddud (2018) further affirmed that with the use of AI driven by big data combined with human behaviour, a new degree of intelligence, innovation, and collaboration could be attained within organisations wishing to achieve digital transformation. It is, therefore, almost practically impossible to describe digitalisation without big data. Che *et al.* (2013) submitted that due to the increase in the use of a variety of digital devices, there is an ever-increasing stream of digital data being generated. This stream of data is today known as big data, on which digital technologies depend to function effectively. For example, Gromov *et al.* (2015) noted that although the main feature of IoT is the connected devices, the intelligence with which this connection is achieved is dependent on the data that the devices are sharing.

Through this data sharing, new knowledge is formed and also possible lines of action are discovered. Because these connected devices generate and share data, it is not unlikely that a large sea of data will be created as a result. These data are bound to grow very big and very fast. Katal *et al.* (2013) noted that big data has the characteristics of the "5Vs". This can be in terms of *variety* as these data are from various sources, types, and formats that are becoming richer by the day. Another characteristic is its *volume* which is enormous as it encompasses the amount of data gathered, stored, and calculated. The *velocity* at which these data are processed is also a unique characteristic. Big data are assessed and processed faster to assess high-value information for different types of data. The *value* of information gathered through this fast speed is very high, while *variability* in data gathered is assured due to increase in the use of diverse communication platforms. Gromov *et al.* (2015) further noted that as individual pieces, data generated are meaningless, but when put together in a mass can generate a vast wealth of knowledge. Herein lies the concept of big data.

According to Rüßmann *et al.* (2015), big data can be described as gathering data from diverse sources and their comprehensive evaluation to provide strategic information that will support real-time decision-making. Bagheri *et al.* (2015) submitted that the analysis of previously recorded data could be used to determine the issues in the previous process and forecast future challenges that might arise and strategies towards mitigating these issues. Che *et al.* (2013) mentioned the importance of big data in the future of all aspects of human life. It was stated that even governments in developed and some developing nations now see the importance of mining big data from social media and blogs, online transactions and other sources of information to determine areas where government facilities are needed, recognising suspicious groups that can pose threats to the society and even predict both positive and negative future events.

In construction, Ahiaga-Dagbui and Smith (2013) noted that the industry worldwide is currently experiencing "data boom" due to the high demand for construction products and technology advancements being introduced into the delivery of these projects. Large data generation is difficult to handle using the conventional data analysis approach that the construction industry has become accustomed to. However, proper analysis of these data could significantly improve how construction industry delivers its services. This has led to digital data mining within the construction industry (Ahiaga-Dagbui & Smith, 2013; Yu & Lin, 2006). Thus, there is a strong relationship between big data and data mining which involves deep digging into data to get possible trends.

Robotics and automation

The application of robotics and automation in construction is broad. It can be implemented from the planning and design phase through to execution and maintenance/monitoring of the project and ultimately, its demolition (Strukova & Liska, 2012). However, despite being one of the oldest industries with a high impact on the economic sector, the construction industry has been inactive in research and development (R&D) for robotics and automation. Robotics and automation research in the industry began in only recent times, with more interest in service robots (Balaguer & Abderrahim, 2008). Making a case for the use of robotics and automation in construction, Cai *et al.* (2018) stated that due to the ageing population in most countries around the world, there is a massive decrease in the number of young workers in construction sites. This decrease is also due to lack of safety and fear associated with construction, particularly in high-rise buildings. Previous studies have also emphasised construction work's danger as the industry records the highest occurrences of injuries and fatalities every year (Brace *et al.*, 2005; Lipscomb *et al.*, 2006). This lack of labour and non-assurance of the few available workers' safety has led to the apparent need to leverage machines and robots to replace human effort in construction (Cai *et al.*, 2018).

Industrial robotics and automation have been described as a critical driver of 4IR. This is due to continuous improvement that has made robots more autonomous, flexible, and cooperative. With this development, certainly, robots will soon be able to interact with one another and work safely side by side with humans and learn from them, and this is the goal of 4IR (Bahrin *et al.*, 2016; Rüßmann *et al.*, 2015). The use of this technology in construction will help avoid site fatalities that tend to occur due to the hazardous nature of construction sites. Human involvement is reduced with robots, and so is fatality issue (Bahrin *et al.*, 2016). Also, robots' use gives more precise autonomous production as tasks are completed intelligently within the given time frame while ensuring safety, flexibility, and versatility (Bahrin *et al.*, 2016). Although some see this technological advancement as a problem for construction workers due to the fear of robots displacing workers on site (Oke *et al.*, 2017), its adoption will allow the effective use of workers for the tasks that are more important for which they can be more productive (Bahrin *et al.*, 2016).

Today, new robotics systems are developed to automate current machines for the automation of construction processes. Although this technology is still at its infancy stage (Aghimien, Aigbavboa, Oke, & Thwala, 2019), its adoption is seen in the aspect of bricklaying, internal finishes, and prefabricated construction. Furthermore, its usage can be on-site or off-site construction using construction robots such as bulldozers and dump trucks, backhoes, and fully automatic controlled shield machines (Yahya *et al.*, 2019). In prefabrication of building components, robotics and automation are high in precast concrete production, shifting from mass production to mass customisation. This development was made possible by using robotic cells to carry out several tasks, such as placing reinforcement bars and setting moulds (Bock, 2008).

The slow adoption of robotics and automation in the construction industry has been attributed to several challenges, and the unstructured nature of the construction environment is one of them (Balaguer & Abderrahim, 2008). The construction environment was described as one that *"involves handling heavy objects, elements made with big tolerances, low level of standardisation, medium level of industrialisation and prefabrication, in addition to the intervention of numerous non-coordinated actors (architects, builders, suppliers, etc.)"*. Pereira *et al.* (2002) in a study on the use of robots on construction sites submitted that robots will only be practically possible in a factory where prefabrication of building components is carried out than on an actual construction site where movements need to be controlled. This submission has been proven somewhat inaccurate as the use of robots and autonomous vehicles are now evident on actual construction sites as in the Crossrail project in London (Castagnino *et al.*, 2016).

Drawing from studies conducted on the use of robotics and automation in manufacturing and construction (Cai *et al.*, 2018; Tambi *et al.*, 2014; Vähä *et al.*, 2013), an important lesson to learn is the fact that the choice to use robotics and automation on a project must be decided from the early design stage of the project. Plans must be made to use it all through from start to completion. Proper structuring of the environment and controlling of same is essential. This is because this technology will influence the design of the building, the planning and organisation of the processes, the methods to be adopted, and the process of construction itself. Through this predetermination of the use of robotics and automation from the onset, its actual usage can be easier as all necessary factors must have been considered and precaution for its successful usage must have been put in place (Kim *et al.*, 2015; Taylor *et al.*, 2003).

Three-dimensional printing

Three-dimensional (3D) printing, also known as additive manufacturing, is an automated process that requires the use of software, hardware, and materials, all working together to produce a reliable final product. Various 3D software programs in the construction market can be used to create a 3D digital model. Once the digital model has been created, it is then scanned with a 3D scanner, and the scanner processes the digital model into a readable file format for which the 3D printer can process. The 3D printer then extrudes printable material on-site in accordance with the layers and processes of the 3D digital model (Sakin & Kiroglu, 2017). This process has been described as the act of creating a physical object from a 3D digital model in a layer-by-layer process (Lim *et al.*, 2012).

The most common technology deployed in the construction industry for 3D printed structures is called Contour crafting. Contour crafting can produce solid structures that are greater than one metre and involves extruding cement base paste from the 3D printer nozzle to form the internal and external skeleton walls of a building. The walls' inside is filled with a bulk compound like concrete (Wu *et al.*, 2016). Contour crafting uses two basic methods for producing the 3D object. It uses Fused Deposition Modeling (FDM) and layers until the whole product is complete. FDM requires ductile materials that self-harden during the cooling process to be extruded by the 3D printer's first nozzle and the second nozzle's deposition. The two nozzles are in the printer head of the robotic arm. The robotic arm of the 3D printer deposits the ductile materials from its nozzles by following the cartesian coordinates set out in the 3D digital model, hence creating cross-sectional layers (Hager *et al.*, 2016).

Aside from Contour crafting, the Batiprint 3DTM method, like the Contour crafting, has also been used directly on site. The Batiprint 3DTM is a method currently developed as an alternative to contour crafting. It requires deposition of polyurethane foam into parallel layers to form the formwork of building walls. Once the polyurethane is deposited, it expands and stiffens rapidly. After a slight delay, self-compacting concrete is printed between two polyurethane layers. Another off-site 3D printing process that has been adopted over time is the D-shape Monolite which uses the layering of powdered substance in a powered bed until the required depth and thickness have been met. Compaction and binding are then applied to the intended solid parts; afterwards, the solid structure can be dug out of the powdered bed (Lim *et al.*, 2012).

The first 3D printer was created in 1984, and over the past 20 years, developments on 3D printers have been kept secret in engineering laboratories. 3D printing is currently a fastest-growing technology across different major industries globally (Hager *et al.*, 2016). While the construction industry has been noted to have stayed the same as 100 years ago in terms of building techniques used and materials, developments of commercialising 3D printing in the construction industry are becoming popular, hence increasing the potential to change the construction industry's landscape (Sakin & Kiroglu, 2017). The number of new entrants into the 3D printing market is rising. In 2013, it was recorded that there were approximately only 20 start-ups in the 3D printing construction field. Current reports list about 65 business entities that offer services including prototyping solutions, software and design tools, 3D printing of large components, and 3D printing of entire buildings. Technological advancements in 3D printer equipment and materials are continually improving, and their costs are falling. Developers in the market are continually creating new devices or printing methods. In China, 3D printing technology was utilised to construct ten one-storey houses within 24 hours (Goldin, 2014). In Dubai, the "office of the future" was constructed within 19 days; 17 days to print building components and two days to assemble the structure on site. The building was initially used as a meeting space. The 3D printed office is a fully operating structure with services like electricity, water, telecommunications, and air-conditioning systems (MacRae, 2016).

The use of 3D printing promises immense benefits for the construction industry. Among these benefits are rapid prototyping and mass customisation, reduction in labour and material costs, reduction in transportation costs and required storage areas, reduction in site injuries and fatalities, increased design creativity, creation of new industries and new jobs, and improved productivity and performance in the construction industry (Fonseca, 2018; Hager *et al.*, 2016; Labonnote *et al.*, 2016; Lim *et al.*, 2012; Pîrjan & Petroşanu, 2013; Sakin & Kiroglu, 2017; Mohd-Tobi *et al.*, 2018; Wu *et al.*, 2016). While these benefits exist, the full adoption of this technology is also challenged by factors such as lack of awareness of 3D printing methods and benefits, lack of government policies or support, lack of practical adoption and research on the real implementation of 3D printing construction methods, high costs of 3D printer, additional costs of secondary devices used alongside the printer, limitations of cementitious printing material, unfavourable existing building regulations, construction contracts labour requirements which require the employment of low and unskilled workforce that may not necessarily be needed by 3D printing approach, and the need for reskilling of the workforce (Berman, 2012; Dickinson, 2018; Hager *et al.*, 2016; Jin & Yu, 2016; Lanko, 2018; Lim *et al.*, 2012; Mohd-Tobi *et al.*, 2018; Paul *et al.*, 2018; Perrot *et al.*, 2016; Sakin & Kiroglu, 2017).

Augmented/virtual realities

Virtual reality (VR) and augmented reality (AR) are two fields mostly used interchangeably and perceived to mean the same thing. AR is essentially the process of adding virtual or imaginary objects onto existing ones to create a feel as to how these objects will relate/fit into the pre-existing environment. AR allows the user to view a real-world scene by creating an environment with computer-generated information (Chi *et al.*, 2013; Raajan *et al.*, 2012). On the other hand, VR is fictional as it stimulates an entirely non-existent atmosphere. VR simulation creates a 3D image of an environment that surrounds the user in which the user will have an insight into the real world and how it operates (Li, Yi *et al.*, 2018). Raajan *et al.* (2012) explain that AR combines simulated components with actual objects and is interactive, which gives one the feeling that an object is there when it is not. It provides information to construction workers about their ever-changing work environment without overloading them with information.

AR systems can produce online images of the construction amplified with the rendering of Computer-Aided Design (CAD). Moreover, 3D pictures or 3D scans help in creating geometric models of the existing structure or site. These models can be compared with the design models to detect problems. This is beneficial as a computer-aided construction inspection system can be built, which will positively affect productivity by circumventing construction errors. Likewise, AR can be used for effective construction project programming, progress tracking, skills transfer, safety compliance, time and cost management, quality and defects management (Ahmed *et al.*, 2017; Kwon *et al.*, 2014).

Furthermore, as the construction industry is a high-risk industry with a high rate of accidents, VR has been developed to improve safety by providing a construction safety training practice. For instance, a virtual platform has been developed to enable university students to access safety information by scanning QR codes using mobile devices. VR can also be applied in equipment and operational task training through simulation-based training (Wang *et al.*, 2018). Construction excavators can use a virtual training system which is based on a game tool engine for the training of workers. Moreover, wearable devices exist to help engineers access on-site information instead of taking construction drawings to site as the device presents the drawing and related information. Developed AR planner system allows the construction worksite planner to locate and arrange construction materials and equipment using a representative 3D model object in the virtual world in the planned worksite (Kodeboyina & Varghese, 2016). Research has revealed that most workers chose mobile-based AR instead of manual capturing of site information for progress reports as they perceived this method to be quicker and easier. Both tools also create a network that enables conferences with multidisciplinary consultants and other project participants who are geographically dispersed (Gheisari *et al.*, 2014).

Cloud computing

Cloud computing is "*an extensive distributed computing standard whereby a pool of virtualised, scalable, vastly available and manageable computing resources (e.g., networks, servers, storage, software, hardware, applications, data) could be achieved, utilised and managed with minimal effort*" (Rawai *et al.*, 2013). Sage (2012) defines it as the storage of data outside of the firewall of a company through virtual servers that can be accessed through the internet without introducing additional IT infrastructure within a company. Similarly, Kumar and Cheng (2010) defined cloud computing as "*a general term for anything that involves delivering hosted services over the Internet*". Cloud computing is regarded as one of the essential elements that form the cyber-physical systems (CPS), and it is expected to play a significant role in 4IR. In this case, cloud computing allows centralised, shared, and variable expansion of computing resources by prompting a new race of global IT developments (Atobishi *et al.*, 2018).

Cloud computing technology is still emerging in the construction industry as big organisations have security concerns regarding its adoption (Beach *et al.*, 2013). The primary purpose of cloud computing is to enable multiple users to access stored data and computation without each needing to use individual licence (Ismail *et al.*, 2018). Information can always be stored in a "cloud" and accessed via the internet or extranet at a later stage for reference and statistical purposes (Mandičák *et al.*, 2016). This ease of storage and access provides means of detecting methods and decisions that worked or did not work without contacting previous personnel who have retired or without searching through paper-based archives. Furthermore, with the continual change of new site locations and workers as evident in construction, cloud computing in the construction industry becomes apparent. Workers need to have better access to the company data to aid in timely decision-making and send reports while working on site (Sage, 2012).

Cloud technology has opened new prospects that enable users to make instant connections and tap into back-office data securely from any location where there is access to the internet (Sage, 2012). With construction where site meetings and client's meetings require physical presence, cloud technology aids to balance the function of back offices such as logistics planning, producing financial reports, payrolls and billing. Using this technology, employees can access the work information in any location without having decentralisation of offices (Rawai *et al.*, 2013). This minimises a lag in information distribution, saves time wasted on scheduling on-site meetings, enhances effective collaboration, and saves cost and time (Chan & Liu, 2007). Cloud computing can also be used for various functions within the construction industry such as the storage of architectural schematics, structural analysis, cost estimating, programme scheduling and control, and procurement, among others (Kumar & Cheng, 2010).

Artificial intelligence

Newton (2018) defines artificial intelligence (AI) *"as being the ability of computer systems to perform tasks normally requiring intelligent human intervention"*. AI captures large amounts of data and analyses the information for patterns and trends. It uses machine power to model human intelligence and machine learning to solve problems in a faster and precise manner (Clavero, 2018). Due to the human–machine relationship involved in the use of AI, human–AI trust governed by appropriate legislation and regulations is believed to be a crucial requirement for the successful implementation of this advanced technology in any organisation (Parveen, 2018; Schia *et al.*, 2019). While not much has surfaced in the use of this technology in construction, AI can revolutionise the construction industry, much like all other technological tools that have been invented. Moreover, particular importance is its ability to provide real-time data about a building regarding structural and maintenance issues; this will enable planned maintenance. On the other hand, it will make it simpler for designers and engineers to improve designs based on historical records stored digitally (Newton, 2018).

Blockchain technology

Blockchain, an advancement in distributed ledger technology, is increasingly being investigated as one of the solutions to some of the challenges faced by the construction industry, which include low productivity, poor compliance and regulation, lack of collaboration and information sharing, and poor payment practices (Li & Kassem, 2019; San *et al.*, 2019). Blockchain technology can be described as a distributed digital ledger decentralised and shared among different networking entities mostly described as nodes. These entities can be individuals or companies, and the blockchain system can serve as a peer-to-peer value transaction system with no need for transactions' in-between verification, security, and settlement through trusted intermediate third-parties (Kifokeris & Koch, 2019). Blockchain is regarded as a game changer in many industries and is expected to yield the same results in the construction industry. Blockchain technology aims to transform commercial space and lead global economies (Tapscott, 2016). The technology also offers transparent solutions which operate securely both privately and publicly, and it is characterised as the "internet of value" (O'Boyle, 2017; Tapscott, 2016). Similarly, using cryptocurrency, blockchain ensures security and anonymity through a distributed tolerance consensus mechanism (Kypriotaki, 2015). In the aspect of construction supply chain management, blockchain technology offers a solution to the tracking of data, sharing of data, as well as providing smart contracts (Qian & Papadonikolaki, 2019).

The blockchain's importance is gained through the implementation of smart contracts. These self-executing pieces of code execute the contract's terms once preset obligations are met (Boucher *et al.*, 2017). Since human input and control may still be needed, smart contracts may be considered automatable traditional contracts. Simply put, smart contracts reduce the need for intermediaries through their if/then commands and the need for physical paperwork is decreased as well (Tapscott, 2016). The Windsor field-rock report further states how smart contracts' incorruptible nature can mitigate risks found in BIM through their hack-resistant nature. When taking on smart contracts, it is imperative to note that some untested legal issues and some unclear contractual terms to mitigate obligations, and dispute risks may exist (Winfield & Rock, 2018). The potential application of blockchain technology in the construction industry includes procurement and supply chain management, payment and project management, and BIM and smart asset management (Penzes, 2018).

Digital twin

Digital twin is one of the emerging technologies and can be referred to as a live model of the physical skill or system which can adapt to changes in the environment or operations. It can illustrate how a product would perform pre-construction (Gopinath *et al.*, 2019). With a digital twin, users can illustrate the state of the past and present physical objects and forecast the future state of these objects (Johnson Controls, 2019). The ability of digital twins to predict future state or processes reduces risks and saves costs (British Standard Institution, 2018). Digital twins are virtual representations of physical elements updated regularly to accurately reflect the real product. It is an extension of BIM and created based on existing technologies and initiatives (*ibid.*). The application of digital twins in the construction industry includes fault predictions on equipment; used on operations giving project managers real-time view; cost reduction through better cost estimates; risk mitigation and security as it can provide a real-time view of the building and its occupants (Johnson Controls, 2019).

Unmanned aerial vehicles (drones)

Unmanned Aerial Vehicles (UAV) also known as drones are "*any aerial vehicle that does not rely on a human onboard operator for flight, either autonomously or remotely operated*" (Golizadeh *et al.*, 2019). Drones have built-in intelligent stabilisation systems that keep them flying and transfer sensors to carry out the required tasks. They are currently changing how construction organisations operate; however, the industry's application is still at an early stage as industry stakeholders are still exploring the benefits of its adoption (Mosly, 2017). However, it is believed that in the construction industry, drones provide an opportunity to easily access facilities that are difficult to access, such as a high rise or complex buildings (Zaychenko et al., 2018). Drones are useful for monitoring of construction activities and progress, surveying earthworks and building inspection, aerial photography and surveillance, security control, enhance health and safety, quality inspection of buildings and infrastructures, quantity take-off and estimation (Golizadeh et al., 2019; Mosly, 2017; Zaychenko et al., 2018). Drones can also be integrated with BIM to achieve 3D modelling data and create innovative technology applications (Golizadeh et al., 2019).

Drivers of digitalisation in construction

Drivers are positive factors that propel the adoption or implementation of a particular element, process, or concept by an individual, organisation, or a larger body. Understanding the drivers of digitalisation in construction is considered necessary for showcasing the need for digitalisation in the construction industry and the road map needed for attaining digital transformation.

Some critical drivers that influence the espousal of construction technologies can be seen from the environmental and organisational perspectives. Market and environmental factors can significantly drive the adoption of technologies in delivering a construction project. For example, the level of competition in the market, organisation reputation, and loyalty to clients are factors that drive companies to adopt innovative technologies to sustain competitive advantage (Nikkas et al., 2007). From the perspective of BIM usage, it has been observed that pressure arising from clients and competitors has a considerable influence on BIM adoption, with the construction industry being known to be highly competitive and clients requiring contractors to show their capability through a past performance where the contractor has successfully managed BIM projects (Eadie et al., 2013). Similarly, organisational drivers refer to the apparent advantages of adopting a technological innovation in connection to an organisation's specific context (Nikkas et al., 2007).

For technologies to be adopted, there must be well-trained staff to handle these technologies. This underscores the fact that available training for personnel can drive the use of specific digital technologies. These training can be organised by vendors of these technologies at their training facilities to teach users the required skills needed for these digital tools. Therefore, the management of construction firms can take advantage of these training pieces to groom their staff to use the required digital tools (Becerik, 2004). Government support through policies and legislation has also been recognised as a crucial driver for construction organisations in adopting innovative concepts and technologies (Babatunde et al., 2018). Furthermore, the creation of soft loans by government for technological development can help encourage these construction firms to adopt digital technologies. This will considerably reduce the financial burden associated with the acquiring and maintaining these technologies and training of staff (Oke et al., 2018).

KPMG (2016) reported some specific drivers for innovation. These drivers can also be seen as critical to the adoption of digitalisation within construction organisations. They include client needs, meeting demand; competition and market forces, new markets, growth and profitability; efficiency, planning and cost reduction; technology and talent; and increasing regulation. From the perspective of PwC (2016), these drivers include digitisation and integration of vertical and horizontal value chains; digitisation of product and service offerings; and digital business models and customer access. Digitisation and integration of vertical and horizontal value chains are seen from developing the products and purchasing, through manufacturing, logistics, and service. The data received regarding performance procedures, the process of achieving maximum productivity with minimum waste and expense and quality management, and planning operations are accessible in real time, with augmented reality and enhanced in a composite network. The internal performance of products from the end-users and suppliers and all-important value chain partners is what horizontal integration stretches beyond. On the other hand, digitisation of product and service offerings is the addition of the current products. For example, by attaching smart sensors to products or devices that can communicate to utilise them in conjunction with data analytics tools, and creating new digitised products that are entirely focused on integrated solutions by incorporating new procedure or process to collect and analyse data, organisations will have the means to produce data on product use by customers and also improve their products to fulfil the customer's requirements. Lastly, with digital business models and customer access, disruptive digital business models are usually aimed at creating more revenues on digitalisation and enhancing interaction with end-users and improving access to end-users. Digital products and services are often intended to give end-users complete solutions in a distinct digital ecosystem.

Kagermann (2014) observed some critical drivers for proper digital uptake in the manufacturing industry. Considering the similarity between the manufacturing and construction industries where production to consumption is the key, these same drivers can assist digital adoption within the construction industry. These drivers include standardisation, process/work organisation, product availability, adopting new business models, having the right security in terms of technical know-how, having the right set of staff, research and development (R&D), training of employees, and availability of regulatory framework. Berger (2016) submitted four key aspects which are necessary for digital transformation, and they are digital data which involves electronic collection and analysis of data; automation which involves the use of new technologies to create autonomous, self-organising systems; digital access which involves mobile access to the internet and internal networks; and connectivity that has to do with connection and synchronisation of hitherto separate activities.

Dimick (2014) observed how organisations could improve their productivity and conveniently adopt digital technologies. First, organisations are advised to know their processes and keep it simple. It is essential to identify the organisations' processes and work assiduously to support those processes as simple and straightforward as possible. By thoroughly knowing the processes, it will be easier to assess at the outset the organisation's current and future needs in terms of digital technologies. The second step is the full commitment of management within the organisation. However, even though total commitment throughout the organisation may be initially unachievable, key players' clear and visible commitment from the outset is essential. Third is the need to enlist the right set of people for the job. Getting project managers and workers digitally oriented will go a long way in promoting digital technology adoption within the organisation. Next is having a defined end goal; a clear vision of the adoption of digital technology. Keeping this vision in mind helps ensure that it remains the organisation's priority and helps avoid scope creep. Lastly, clearly defining the organisations' business process, garnering management commitment, getting the right set of people and having a clear vision, it is important to plan the actual adoption process carefully and stick to the plan. Based on the preceding, the drivers of construction digitalisation can be summarised to include the following:

- Pressure from clients and competitors
- Emerging markets that require the digital capability
- Availability of required digital technologies
- Availability and easy access to data and network connectivity
- Availability of technical expertise to handle digital technologies
- Availability of required digital training and education
- Government support through policies, legislations, and provision of financial assistance
- Standardisation of construction works
- Adoption of new business models
- Effective R&D
- Need for growth and profitability within organisations
- Need for efficiency, planning, and cost reduction in project delivery

Barriers to digitalisation in construction

Barriers are seen as negative factors that act as obstructions to adopting or implementing a particular element, process, or concept by an individual, organisation, or a larger body. These factors are also referred to as "challenges", "impediments", "obstructions", "hindrances", or "factors affecting" a particular concept under investigation. In this section, some of the factors that could act as barriers towards the proper adoption and implementation of digital technologies in the construction service delivery are explored.

One of the significant issues affecting the appropriate utilisation of technological innovations in construction is the absence of well-trained technological staff. The absence of right talent needed to attain digital transformations is one key hurdle hindering digital transformation in most construction organisations. To get the required talent, organisations may have to acquire new talent, train existing employees, or collaborate with organisations with the required talents (Sacks &

Barak, 2010). Oke *et al.* (2018) observed that in most developing countries, there is no specific module designed towards digital training in the majority of institutes of higher learning, and this harms the construction industry. This is because construction firms have to train their graduates further to be digitally learned. To a large extent, this can serve as a source of discouragement for construction firms who are already struggling to survive within the competitive construction environment.

The issue of training workers on usage of digital technologies depends on the financial capability of the organisation. In most developing countries where construction firms are mostly small, medium, and micro enterprises, training of workers in the use of innovations might prove difficult. Rather than training, these firms stick to old practices, leading to poor service delivery and poor competitive edge. The high costs involved in acquiring and maintaining digital technologies such as robotics and automation, UAVs, and BIM are among the barriers of construction digitalisation (Golizadeh *et al.*, 2019; Yahya *et al.*, 2019). Similarly, the high cost for R&D innovation, as it requires a high qualified workplace as well as an increase in company's capital intensity and the high cost for updating current technologies, costs for training employees in using the technologies and cost involved in changing construction operations are all financial challenges for construction organisations seeking to be digitally transformed (Vaduva-Sahhanoglu *et al.*, 2016). Dimick (2014) and Bédard-Maltais (2017) observed that the financial investment needed to fund technology adoption is often considered analogous to adding another salary to the payroll. It was observed that most organisations would prefer to hire more personnel rather than acquiring new digital technology because of the associated cost. This view can be myopic, as these organisations are not considering the accuracy, speed, and productivity associated with the use of digital technology.

Aside from finance, security and safety of data and information have also been noted as a crucial barrier to construction digitalisation. Considering that most organisations' activities will be done via the internet, data security and protection are essential for organisations' successful day-to-day running. Organisations need to be assured that information and data are open to only the right individuals. No doubt, computer security and protection procedures borders on the safety of important parts of information, resources, data, and assets that incorporate intellectual properties in these pervasive and frequently open information eras (Zeng *et al.*, 2012). Therefore, it is imperative that access to information is only available to and by the right individual at the right time. Looking as UAVs, these drones capture a large amount of data which could sometimes include confidential or sensitive information of private properties or private behaviour. This makes privacy a crucial problem for organisations seeking to adopt this technology (PwC, 2016).

Similarly, with the use of IoT, users are concerned about the privacy of their devices and data as IoT enables inter-communication between connected devices with each other and with humans. Furthermore, there is the concern about daily personal, home, and work devices being pervasively exposed to the vast network of interconnected smart devices (Rad & Ahmad, 2017). A similar observation was made in the use of BIM and cloud computing. While BIM involves information sharing, which allows the project team to access project data, this can cause copyright infringement and unauthorised access. In the same vein, security concerns such as phishing and data loss are the main challenges related to cloud computing adoption (Atobishi, 2018)

There is also interoperability, which is the capacity of organisations and professions to trade, offer, or incorporate data and business forms, crosswise over data frameworks or authoritative practice. When this is missing, there is a tendency to fail diverse frameworks to connect adequately (Eastman *et al.*, 2011). This practice can be overwhelming as a result of the degree of fracture, differing qualities in data prerequisites, the inflexibility of data schemas (standards), requesting innovation competency necessities, and more extensive business goals, which anticipate merchants' compatibility in the improvement of programming and frameworks that interoperates (Doller & Hegedorn, 2008; Isikdag *et al.*, 2007).

Some studies have also asserted that resistance to change can be a crucial problem hindering digital technology adoption (Strukova & Liska, 2012; Vaduva-Sahhanoglu *et al.*, 2016). Most construction workers and unions do not readily accept the technologies as they view these technologies to replace workers. Additionally, the lack of commitment by the public sector clients and resistance in changing the current work practices hinders digital technologies' adoption (Oke *et al.*, 2018; Yahya *et al.*, 2019). Dimick (2014) opined that some level of distrust in new technologies exists in most organisations. Change can be challenging for humans, and in most cases, organisations tend to fall victim to this challenge. This resistance cuts across every level of these organisations, from workers to executive decision-makers.

The technical difficulty arising from the use of digital technologies such as UAVs can serve as severe hindrances to this technology. This difficulty can be in the form of a large volume of the data generated and data loss, lack of accuracy in detecting on-site dynamics and failure of global position system (GPS) signals as well as the site-related problem that interferes with project activities and obstacles on construction sites (Golizadeh *et al.*, 2019). Furthermore, restrictive regulations and lack of information on laws and regulations hinder the implementation of some of these technologies in the construction industry (Golizadeh *et al.*, 2019; Mosly, 2017). Also, uncertainty on insurance coverage availability covers for risk of physical loss or liability during and after the operation. This is because laws surrounding the use of these technologies are still evolving in most countries worldwide (PwC, 2016). In the same vein, legal issues such as liability arising from the involvement of multiple stakeholders contributing to the model in BIM adoption can be a severe issue. Intellectual property rights and model ownership as BIM models can easily be copied and extracted can pose a severe challenge to technology usage (Sardroud *et al.*, 2018). Also, the contractual issue of who will take responsibility in entering and updating data into the model and taking accountability should there be inaccuracies is equally a barrier to adopting such digital technology.

Based on the preceding, the barriers to construction digitalisation can be summarised to include the following:

- Lack of technical expertise
- The high cost of acquiring and maintaining digital technologies
- The high cost of training skilled personnel to handle digital technologies
- Lack of education and training in the use of digital technologies
- Size of the organisation
- The poor and high cost of R&D in the construction industry
- Technological difficulties/complexity of technologies
- Construction industry's resistance to change
- Legal concerns
- Lack of government support
- Data security and privacy issues
- Lack of awareness and demand
- Poor management commitment and support
- Problem of interoperability and incompatibilities of systems
- Poor infrastructure (power and internet connectivity)
- Quality control issues
- Lack of convincing business evaluation
- The poor digital culture within the construction industry
- Need to restructure the organisation's activities and structure
- Poor standardisation
- Client and managements' worry on profitability

Risk associated with digitalisation in construction

Despite the number of benefits inherent in using digital technologies in transforming construction activities and offering better services to clients, the uptake of these technologies in the industry is still slow. In light of this, it becomes necessary to collate information regarding the negative aftermaths (risks) of digitalisation to put investors and clients at ease by giving them the platform to understand the risk and evaluate whether they can be managed, transferred, or avoided. In support, Cobb (2001) mentioned that to effectively implement digital technologies such as robotics and automation in the construction industry, there is the need to evaluate the risk involved early stage before implementation. It was further noted that most contractors have problems with the fact that digital technology vendors provide information relating to what the technology can do, not what it is unable to do, and potential weaknesses and implementation issues.

Based on the aforementioned, the likely risk factors that might occur due to the adoption of digital technologies in construction organisations are x-rayed in this section. It has been noted that the construction industry is known for its risk-averse nature (Hudson, 2017). This notion deems it appropriate to define the term "risk". Risk is the likelihood of occurrence of an undesirable event. It can also be defined as "*the actual exposure of something of human value to a hazard and is often regarded as the combination of probability and loss*" (Gravley, 2001). Through early determination, the amount of risk can be controlled and lessened.

Financial risk

The construction industry is conventional, and big firms remain reluctant to allocate substantial funds in the investment of digital technologies because there is insufficient data that promises returns on investment in the immediate future (Yaghoubi *et al.*, 2012). Technological tools need to be adopted and, more importantly, implemented for investors to reap all the promised benefits. However, it is essential to note the difficulty in predicting annual cash-flows relating to adopting these digital tools (Taylor & Smith, 2000). It is necessary to highlight the possibility of insufficient returns on investment due to technology being successfully implemented but not used to its optimum function. Moreover, digital technology broadens the market through industry competition as it gives access to the market, individuals, and companies which could not previously participate. The consequence of this is that profit margins will drop. More critically, cheap labour availability should not be ignored as an influencer of whether to digitalise. Digital transformation may be halted by comparing labour costs matched against the high price of technological tools. Moreover, ancillary labour factors such as increased protests and unionisation might influence a digitalisation decision despite the exorbitant cost of investment in digital gadgets and applications (Le Roux, 2018; Meno, 2020).

Productivity loss

The increased use of computers and access to the internet open up the temptation to indulge in unproductive activities during working hours. This is a reduction in time spent on actual work tasks (Gaille, 2016; Love *et al.*, 2001). Furthermore, the overreliance on computers will result in work stoppages in instances where there is loss of power as experienced in most developing countries particularly in Africa or during a sudden computer system crash, and all the information needed to work is stored on the computer (Gaille, 2016).

Uncertainty of whether the tools adopted will meet expectations

The possibility that the adopted digital technology will not produce the desired result will invariably lead to disappointment for the technology adopter. Stephenson and Blaza (2001) observed that implementation issues might occur with introducing new technologies within an organisation. For example, with radio-frequency identification (RFID), it is noted by Kopsida *et al.* (2015) that some construction elements cannot be fitted with tags; this requires extra costs concerning additional labour-complementary equipment to supplement this. The successful deployment of technological tools depends on more than the straightforward purchase or investment in the tools but robust and effective leadership and planning:

this needs to be understood. Based on the construction industry's intrusive nature into other sectors that govern its successful operation, the adoption of new technologies has the potential of creating material conflict among these different organisations and occupation groups, where technology might in some areas make processes simple for one organisation but create an administrative nightmare for another organisation. This creates power struggles in the process (Meno, 2020).

Adoption of the technology in silos

The engineering and architectural fields are the highest adopters of design technology such as BIM (Young *et al.*, 2008). This is an isolated group among the various professions found in the construction industry. What happens in most construction firms is that new technology is adopted in silos and people are unwilling to share information about what the technology can do and how to use it (Kane *et al.*, 2015). It is not uncommon to find only one planner in the construction firm who can use the application or device. This creates an issue once the individual is inundated with work or if the individual leaves the company, and training has to be administered from scratch.

Security

A security risk may arise from cyber attached or asset management. As the construction industry progresses and becomes more dependent on digital tools for remote working, communication, and the storage and quick access to data, the threat from cyberattacks will also increase (Hudson, 2017). A significant issue relating to digitalisation is cybersecurity as information stored in the cloud network system can be hacked by malicious IT hackers if not adequately protected (Rubin, 2006). This is a concern for many reasons such as architectural and engineering design copyrights (intellectual property) and leaks of confidential rates in tender pricing. Cybersecurity is a threat that has not been given much attention to the built environment (Pärn & Edwards, 2019). Furthermore, theft on site is not a rare occurrence; moreover, break-ins are also prevalent, especially where the project takes place within a residential environment. This results in several gadgets that will need replacing (Ahsan *et al.*, 2007).

Job losses/change in the job market

When software was introduced in the market to render two-dimensional illustrations using CAD, this era resulted in manual draftsmanship becoming obsolete, and people who possessed this skill became redundant. Digitalisation in the current period will have the same effect with the deployment of software such as BIM, should it be widely adopted (Watson, 2011). However, this submission by Watson has been disputed with the submission that a technological shift into work process will not necessarily replace jobs but will introduce a new demand for different skill sets (Lawrence *et al.*, 2017; Muro, 2017). This highlights the importance of voluntary learning and individual interest and effort in searching for new skills (Lawrence *et al.*, 2017). However, it has been observed that digitalisation has the potential to open the market to skills outsourcing. In this case, local skills will be at a disadvantage since hiring locally will be more expensive. Therefore, technology adoption has favourable business results through increased productivity and efficiency but bodes adverse effects for the workforce. Digitalisation will also lead to employers reducing their workforce since fewer human resources will be required to conduct essential duties. Over the years it has been observed that some technology has replaced blue-collar jobs. As digital tools become more intelligent, what will be observed is an increase in the replacement of white-collar jobs (Le Roux, 2018).

Digitalisation foresees a shift in the human resource structure as well. In a theoretical sense, pay scales will not necessarily be proportional to education level and experience but rather to tech-skills (Lawrence *et al.*, 2017: Mahbub, 2008). The more senior employees are currently paid higher wages for their experience: this is mostly based on conventional methods of working. The adoption of digital technology will mean that prospective employees will face tech-skills as a prerequisite for employment and attract such skills; employers will have to adjust their remuneration scales (Le Roux, 2018). Therefore, it will cost more initially but become more beneficial to hire technologically skilled personnel than to employ a senior candidate who has to resort to conventional methods in a company that is intent on digital transition. Mahbub (2008) also finds that low- to medium-level jobs that allow individuals with minimal or no educational background to enter the job market will diminish in future owing to the demand for tech-skills. This means that opportunities will move from an education level to the ability to read and analyse digital data, seeing that over two million high-paying jobs do not require a university degree but digital skills.

Misaligned objectives with public sector regulations and policies

The public sector positively influences digitalisation in regions such as the USA, China, and the UK by administering appropriate policies and regulations. This approach might not necessarily work for developing counties where digitalisation might conflict with government strategies. An example is the Expanded Public Works Programme (EPWP) in South Africa, where the main focus is on job creation, emphasising "labour intensity" (CIDB, 2017; Department of Public Works, 2014). All public sector projects prescribe a requirement in the contractor's tender process to commit to spending as much as 30% of the contract value on manual labour. This prescription compels contractors to use minimal machinery on site. More so, the public sector tenders still require rates in the tender bills of quantities to be handwritten, notwithstanding all the tools that can make this process quicker and simpler. Digitalisation might be available to those who can afford and access it. However, to this effect and based on lack of information, it might just as well remain a dormant concept (Meno, 2020).

Legal risks

The construction industry is intensely contractual and document-based with efficient communication being at its core for project success. The legality of documents or any form of information communicated electronically needs to be considered. This is because laws have not yet been completely adjusted to cater to the use of digital tools in the construction industry. Moreover, the issue of intellectual property rights comes into question where a multidisciplinary group of professionals come together to design using a single model: how the ownership of the design data is determined, who provides indemnity, whether everyone is jointly and severally liable for any design failure are the issues to be decided on (Azhar, 2011; Mason, 2017; Meno, 2020). Another point to look into is the frequency of design changes facilitated by each individual, concerning how it potentially affects the rest of the works and the tracking of such changes. The contractor/subcontractor might not keep up with design revisions (Hamdi & Leite, 2014).

In the case of e-business or e-tendering, the debate still lingers around legal documents, where some project participants still regard contracts as agreements done in writing, especially in terms of signatures (Alaghbandrad *et al.*, 2011). Some clients will not accept digital signatures on legal documents which might cause material disputes.

Rejection

Some studies have noted the rejection of a technological transformation within a company, especially by the more senior workers (Alaghbandrad *et al.*, 2011), resulting in financial losses as employees gravitate back to traditional methods and new digital tools and application remain dormant (also bearing in mind that some applications have licences that are paid for every month). This can be a result of various reasons, chief of which are the difficulty of use of the new tools or simply that the more senior workers are uncomfortable with a new way of working and have a perception that technology and the internet make employees lazy. On the other hand, labourers can potentially boycott machines that are seemingly replacing tasks for which they were previously responsible. These tools threaten their job security, and they will resort to manual means of working (Meno, 2020).

Psychological issues

This is a seldom expressed concern about the use of technology in the workplace and privacy infringement. People may generally believe that technology use leaves no room for error. The use of biometric readers and surveillances on site will create an uncomfortable environment and negatively affect occupational health (Meno, 2020).

Loss of personal and interactive relations

With the consideration of social aspects, digitalisation, including the increased use of technology, might have a negative effect on how people interact, resulting from a decrease in physical interaction between them. Likewise, and on a business level, an amplified use of digital tools will promote remote working: what this does is to blur working spaces and working hours (Salento, 2017). Moreover, technology creates a sense of constant availability. This poses a variety of issues such as remuneration ratio for work after hours including the measurement and monitoring thereof; unhealthy interrelations with family members resulting from bringing work home; and loss of relationships and mutual trust among colleagues, clients, and suppliers from lack of physical interactions. If not planned correctly, it will be an unpleasant surprise to digitalisation investors to realise that technology adoption and its integration into business operations is not a transition that only happens on the surface; traditional methods need to be uprooted and changed to meet and comply with new company developments (Datta, 2000).

Personnel injury

Based on safety being a significant concern in the construction industry, one cannot overlook safety hazards that might occur in a world where human and robots interact (Bruckmanna *et al.*, 2016). Automated machinery functions when and as programmed, and if a safety hazard does not fit some predetermined criteria, the robot will not detect it. This is as opposed to natural human instinct where humans are adaptable to new and unprecedented situations.

Increased industry competition

Seeing that technology allows business measures to expand due to the ability to work remotely, business now happens on a global scale where skills are outsourced and expertise is offered at low margins since the market is open to anyone connected. This causes profit margins to drop; moreover, it brings about the need to verify skills and the verification of professional particulars. This takes time and costs money.

Obsolescence/system failures

Technology advancements are occurring rapidly; the number of newly introduced tools might be difficult to keep up with (Strukova & Liska, 2012). Various developers are continually trying to outdo one another or fix existing programmes while creating versions that are a nightmare to keep up with. Besides this fact, applications call for constant system upgrades which require network access and also cost money. Moreover, if these upgrades are not done as needed, applications might crash. Developers seldom highlight this fact which will be a cost factor that creeps in at a later stage. This is particularly an issue with companies that subscribe to hundreds of licences, bearing in mind that a system failure directly affects production, especially when a task is highly dependent on a computer function.

Impact of digitalisation on construction performance

Despite the aforementioned risks of construction digitalisation, the proper deployment of digital tools in construction activities promises significant positive impacts in diverse areas of construction project performance.

Cost and time

The effects that the adoption of digital technologies can have on the overall cost and time of projects are enormous. For instance, less workload per task for an employee, including reduced use of plants such as scaffolding, security systems, and transport machinery can prove to have a significant effect on the reduction of bottom-line costs. Additionally, robotics and automation improve construction programmes; this lessens costs related to time-based preliminaries (Kamaruddin *et al.*, 2016). The use of digital technologies such as BIM help deliver construction projects in time and within budget. With the use of BIM, clashes in designs are detected early in the project. This saves time wasted on the rework of construction works and its associated cost. More so, electronic tendering and procurement platforms save time and money for construction clients and contractors (Berger, 2016).

Quality

It has been observed that digital technologies can help deliver better quality projects that the clients and other project stakeholders can be satisfied with (Kamaruddin *et al.*, 2016; Woodhead *et al.*, 2018). Furthermore, through digitalisation, effective project monitoring of construction projects and plant and equipment can be achieved (Bogue, 2018; McDonald, 2017). This can provide some measure of quality due to strict project monitoring. Also, the use of digital technologies such as 3D printing, with the incorporation of BIM, will motivate designers to create more geometrically appealing structures without difficulty in envisioning the final product or worrying about their build-ability as well as the concern of additional costs from formwork and moulds that will only be used once for a particular project (Buswell, 2007).

Client and stakeholder satisfaction

With the use of digital technology, engineers and designers can offer the client a variety of design options at the click of a button. This is because computers have enabled illustrations to be done more quickly and have also expanded the geometrical scope, including how designers think (Reffat, 2008). It has been noted that the use of robotics and automation can bring about significant satisfaction to the client and other stakeholders as a more durable and precise construction project can be achieved (Vaduva-Sahhanoglu et al., 2016). According to Aghimien (2019), with the implementation of various digital technologies for the different construction processes within the different life cycles of construction phases, the construction industry will be able to minimise uncertainty and promote satisfactory construction processes for its clients and stakeholders.

Moreover, the importance of communication during or leading to a construction project cannot be understated. Digital tools help facilitate efficient collaboration (Oke et al., 2018; Ullah et al., 2019), and interdisciplinary professionals will easily interact and capture and access information in real-time (Issa, 2002). This will enable ad hoc decision-making interactions where projects will be less prone to reworks, including quicker and simpler access to data in real-time for project monitoring and progress tracking (Hudson, 2017). The adoption of digital tools enables an environment where all parties involved (including clients) are kept in the loop and have access to project data including budgets, programmes, and procurement information (Issa, 2002). Thus, digital technology will result in improved storage and access to information, tracking and security, which results in better coordination and subsequently improved customer relations since the communication and feedback to clients will be optimal. This allows ideas to be shared on solving problems without the need to plan and schedule meetings on location (Hlahla, 2013).

Health and safety

The use of robotics and automation can significantly improve health and safety (H&S) delivery of construction projects as risky jobs are done using robots (Oke et al., 2017; Ruggiero et al., 2016). Hager et al. (2016) and Sakin and Kiroglu (2017) also submitted that with the use of technologies such as 3D printers, hazardous works are done by the printer leaving the less hazardous works to be carried out by humans. By doing so, human lives are preserved (Kim & Haas, 2000). Salento (2017) and Hudson (2017) have noted that using wearable devices with embedded sensors can help monitor workers' health issues and detect problems early. This is done through measuring and capturing health data of workers. By detecting issues around workers overworking, stress and subsequent absenteeism early, measures can be put in place to tackle these issues before they affect the overall project outcome.

Construction productivity

Digital technologies can revolutionise the construction industry and revitalise a sector that is perceived as slow and old-fashioned (Hudson, 2017). The adoption of digital tools within the construction industry is done to optimise efficiency and productivity. For instance, through big data analytics, a real-time view of project progress can be achieved (Armstrong & Gilge, 2016) and measures can be put in place to make corrections should issues arise in the process. Furthermore, with mobile devices' use to view building model simulations, an increase in informed decisions and proactive problem-solving can be achieved (Ahsan *et al.*, 2007; Armstrong & Gilge, 2016). The utilisation of robotics in the industry can help increase productivity, speed, and quality (Kamaruddin *et al.*, 2016).

Sustainability

The construction industry is one of the highest users of raw materials with the industry activities having a significant impact on the environment. Khoshnevis (2003) reveals how construction of a typical single house can produce three to seven tonnes of waste. The use of digital technology in construction offers significant economic, environmental, and social sustainability benefits to the industry. For instance, 3D printing promises an emission-free advantage of harmful substances based on its complete electric operation. Moreover, on a general basis, fewer vehicle and less machinery on site means less fuel consumption and a decrease in the depletion of raw materials. This contribution of technology will lead to better environmental sustainability in construction project delivery. Through digitalisation, some form of social and economic sustainability can be attained as more jobs can be created, and upskilling of individuals to take up these created jobs can be achieved. The new industry can also spring forth with the adoption of digital concepts (Fonseca, 2018; Pîrjan & Petroşanu, 2013), thus improving the economy through job creation.

Summary

This chapter presented a description of construction digitalisation. The chapter explored past studies that have been published on digitalisation-related issues within the construction domain and concluded that construction digitalisation had been viewed from the dimension of data and information management; digital technologies in building design, construction, and operation; digital technologies in civil and heavy engineering; digital technologies for sustainable and smart construction; and skills development for digital technology usage. Based on this submission and the review of general definitions of digitalisation, the chapter defined construction digitalisation. One major highlight of the chapter was the discovery of the dearth of research on digitalisation-related issues within the construction industry of developing countries, particularly in Africa where such research is almost non-existent. This shows a rich research area for academics seeking to do impactful research in these countries' construction industry. The chapter gives a rich insight into the different digital technologies that can be adopted within the construction industry to improve service delivery.

Furthermore, the critical drivers, barriers, and impact of digitalisation in the construction industry were highlighted. By understanding the drivers that will propel adoption and the barriers to avoid, construction organisations can better improve their chances of digitalisation. Since construction itself is synonymous to risk, the chapter also gave an insight into the risk construction organisations might face should they decide to adopt digital technologies for their services. This was done to further prepare these organisations for unforeseen issues that might occur. Risk issues surrounding finance, loss of productivity, the uncertainty of the outcomes of the adopted digital tools, adoption of technologies in silos, security, job losses, legal, psychological issues, and the likes must be carefully evaluated with mitigating measures put in place to cushion their effect on the organisation should they occur.

References

Aboushady, A.M. and Elbarkouky, M.M.G. (2015). Overview of building information modeling applications in construction projects. In Raebel, C.H. (Ed.), *AEI 2015: Birth and Life of the Integrated Building*. ASCE Publishers, Milwaukee, WI, 24–27 March, pp. 445–456

Abubakar, M., Ibrahim, Y., Kado, D. and Bala, K. (2014). Contractors' perception of the factors affecting building information modelling (BIM) adoption in the Nigerian construction industry. *International Conference on Computing in Civil and Building Engineering*, Orlando, FL, 23–25 June.

Addy, M., Adinyira, E. and Ayarkwa, J. (2018). Antecedents of building information modelling adoption among quantity surveyors in Ghana: An application of a technology acceptance model. *Journal of Engineering, Design and Technology*, 16(2): 313–326

Aghimien, D.O. (2019). Digitalisation as a veritable option for construction organisations to achieve competitive advantage. *Proceeding of the International Conference on Innovation Technology, Enterprise and Entrepreneurship*, ICITEE, Kingdom of Bahrain, Manama, 24–25 November, pp. 429–436

Aghimien, D.O., Aigbavboa, C.O. and Oke, A.E. (2019). Viewing digitalisation in construction through the lens of past studies. *Advances in ICT in Design, Construction and Management in Architecture, Engineering, Construction and Operations (AECO), Proceedings of the 36th CIB W78 2019 Conference*, Northumbria University, Newcastle, United Kingdom, 18–20 September, pp. 84–93

Aghimien, D.O, Aigbavboa, C.O., Oke, A.E. and Thwala, W.D. (2019). Mapping out research focus for robotics and automation in construction-related studies. *Journal of Engineering Design and Technology*, 18(5): 1063–1079

Ahiaga-Dagbui, D.D. and Smith, S.D. (2013). My cost rennet over: Data mining to reduce construction cost overruns. In Smith, S.D. and Ahiaga-Dagbui, D.D. (Eds.), *Proceedings 29th Annual ARCOM Conference*. Association of Researchers in Construction Management (ARCOM), Nottingham, 2–4 September, pp. 559–568

Ahmed, S., Aopy, M. and Hoque, M. (2017). A brief discussion on augmented reality and virtual reality in construction industry. Department of building engineering and construction management, KUET, Bangladesh. *Journal of System and Management Sciences*, 7(3): 1–33

Ahsan, S., El-Hamalawi, A., Bouchlaghem, D. and Ahman, S. (2007). Mobile technologies for improved collaboration on construction sites. *Architectural Engineering and Design Management*, 3(4): 257–272

Alaghbandrad, A., Nobakht, M., Hossenalipour, M. and Asnaashari, E. (2011). ICT adoption in the Iranian construction industry: Barriers and opportunities. *Proceeding of 28th International Association for Automation and Robotics in Construction*, Seoul, 29 June to 2 July, pp. 280–285

Ammar, M., Russello, G. and Crispo, B. (2018). Internet of things: A survey on the security of IoT frameworks. *Journal of Information Security and Applications*, 38: 8–27

Armstrong, G. and Gilge, C. (2016). *Building a Technology Advantage: Productivity in Construction: Creating a Framework for the Industry to Thrive.* Global Construction Survey and KPMG, San Francisco, CA

Ashton, K. (2009). That internet of things Thing. *RFID Journal*, 22(7): 97–114

Associated General Contractors of America. (2006). *The Contractor' Guide to BIM*, 1st edition. Available at: www.engr.psu.edu/ae/thesis/portfolios/2008/tjs288/Research/AGC_GuideToBIM.pdf [accessed 2–12–2020]

Atobishi, T., Gábor, S.Z. and Podruzsik, S. (2018). Cloud computing and big data in the context of industry 4.0: Opportunities and challenges. *Proceeding of the International Institute of Social and Economic Sciences*, Sevilla, Spain, 5 March

Atzori, L., Lera, A. and Morabito, G. (2010). The internet of things: A survey. *Computer Networks*, 54(15): 2787–2805

Azhar, S. (2011). Building information modeling (BIM): Trends, benefits, risks, and challenges for the AEC industry. *Leadership and Management in Engineering*, 11(3): 241–252

Babatunde, S.O., Ekundayo, D., Babalola, O. and Jimoh, J.A. (2018). Analysis of the drivers and benefits of BIM incorporation into quantity surveying profession: Academia and students' perspectives. *Journal of Engineering, Design and Technology*, 16(5): 750–766

Bagheri, B., Yang, S., Kao, H.A. and Lee, J. (2015). Cyber-physical systems architecture for self-aware machines in industry 4.0 environment. *IFAC Conference*, 38(3): 1622–1627

Bahrin, M.A.K., Othman, M.F., Nor, N.H. and Azli, M.F.T. (2016). Industry 4.0: A review on industrial automation and robotic *Jurnal Teknologi (Sciences and Engineering)*: 137–143

Balaguer, C. and Abderrahim, M. (2008). Trends in robotics and automation in construction. In *Robotics and Automation in Construction*. In-Tech, Croatia, October, pp. 1–20

Beach, T.H., Rana, O.F., Rezgui, Y. and Parashar, M. (2013). Cloud computing for the architecture, engineering & construction sector: Requirements, prototype & experience. *Journal of Cloud Computing: Advances, Systems and Applications*, 2(1): 1–16

Becerik, B. (2004). A review on past, present and future of web-based project management and collaboration tools and their adoption by the US AEC industry. *International Journal of IT Architect, Engineering, Construction*, 2: 233–248

Bédard-Maltais, P. (2017). Industry 4.0: The new industrial revolution – are Canadian manufacturers ready? *BDC Study*. Available at: www.bdc.ca/EN/Documents/analysis . . . / bdc-resume-etude-manufacturing-en

Berger, R. (2016). Digitization in the construction industry. *Think Act*: 1–16

Berman, B. (2012). 3-D printing: The new industrial revolution. *Business Horizons*, 55(2): 155–162

Bhatla, A., Choe, S.Y., Fierro, O. and Leite, F. (2012). Evaluation of accuracy of as-built 3D modeling from photos taken by handheld digital cameras. *Automation in Construction*, 28: 116–127

Bienhaus, F. and Haddud, A. (2018). Procurement 4.0: Factors influencing the digitisation of procurement and supply chains. *Business Process Management Journal*, 24(4): 965–984

Blažun, H., Kokol, P. and Vošner, J. (2015). Research literature production on nursing competences from 1981 till 2012: A bibliometric snapshot. *Nurse Education Today*, 35(5): 673–679

Bock, T. (2008). Construction automation and robotics. In Balaguer, C. and Abderrahim, M. (Eds.), *Robotics and Automation in Construction*. IntechOpen, London

Bogue, R. (2018). What are the prospects for robots in the construction industry? *Industrial Robot*, 45(1): 1–6

Boucher, P., Nascimento, S. and Kritikos, M. (2017). *How Blockchain Technology Could Change Our Lives: In-depth Analysis*. European Parliament Research Service, Brussels, Belgium

Brace, C.L. and Gibb, A.G.F. (2005). A health management process for the construction industry. In Haupt, T. and Smallwood, J. (Eds.), *Rethinking and Revitalizing Construction Safety, Health and Quality*. International Council for Building Research, Studies and Documentation, Port Elizabeth, South Africa

Bruckmanna, T., Mattern, H., Spengler, A., Reicherta, C., Malkwitz, A. and König, M. (2016). Automated construction of masonry buildings using cable-driven parallel robots. *Proceedings of the 33rd International Symposium on Automation and Robotics in Construction*, Auburn, Alabama, July, pp. 332–340

British Standard Institution. (2018). Digital twins for the built environment. Available at: www.bsigroup.com/globalassets/localfiles/en-gb/events/standards-forum–november-2018/digital-twins.pdf [accessed 19–10–2019]

Bui, N., Merschbrock, C. and Munkvold, B.E. (2016). A review of building information modelling for construction in developing countries. *Procedia Engineering*, 164: 487–494

Buswell, R. (2007). Freeform construction: Mega-scale rapid manufacturing for construction. *Automation in Construction Journal*, 16(2): 224–231

Butler, L. and Visser, M.S. (2006). Extending citation analysis to non-source items. *Scientometrics*, 66(2): 327–343

Cai, S., Ma, Z., Skibniewski, M., Guo, J. and Yun, L. (2018). Application of automation and robotics technology in High-Rise building construction: An overview. *35th International Symposium on Automation and Robotics in Construction*. International Association for Automation and Robotics in Construction, Slovak University of Technology, Bratislava, Slovakia, pp. 1–8

Cartlidge, D. (2013). *Quantity Surveyor's Pocket Book*, 2nd edition. Routledge, London and New York

Castagnino, S., Rothballer, C. and Gerbert, P. (2016). *What's the Future of the Construction Industry?* World Economic Forum. Available at: www.weforum.org/agenda/2016/04/building-in-the-fourth-industrial-revolution/ [accessed 19–11–2020]

Chan, E. and Liu, C. (2007). Corporate portals as extranet support for the construction industry in Hong Kong and nearby regions of China. *The Journal of Information Technology in Construction*, 12: 181–192

Che, D., Safran, M. and Peng, Z. (2013). From big data to big data mining: Challenges, issues, and opportunities. In Hong, B., Meng, X., Chen, L., Winiwarter, W. and Song, W. (Eds.), *Database Systems for Advanced Applications. DASFAA 2013. Lecture Notes in Computer Science*. Springer, Berlin, Heidelberg, Vol. 7827

Chi, H.L., Kang, S.C. and Wang, X. (2013). Research trends and opportunities of augmented reality applications in architecture, engineering, and construction. *Automation in Construction*, 33: 116–122

Clavero, J. (2018). Artificial intelligence in construction: The future of construction. Available at: https://esub.com/artificial-intelligence-construction-future-construction/ [accessed 1–2–2019]

Cobb, D. (2001). *Integrating Automation into Construction to Achieve Performance Enhancements*. CIB World Building Congress, Wellington, New Zealand, pp. 1–11

Construction Industry Development Board (CIDB). (2017). *Construction Monitor; Employment*. Third Quarter, CIDB Pretoria, South Africa, October, pp. 1–20

Crampton, N. (2016). What is digitalisation and what does it mean for your business? Available at: https://bizconnect.standardbank.co.za/sector-news/construction/the-digitalisation-of-the-construction-industry.aspx [accessed 15–8–2018]

Crnjac, M., Veža, I. and Banduka, N. (2017). From concept to the introduction of industry 4.0. *International Journal of Industrial Engineering and Management (IJIEM)*, 8(1): 21–30

Dainty, A., Leiringer, R., Fernie, S. and Harty, C. (2017). BIM and the small construction firm: A critical perspective. *Building Research and Information*, 45(6): 696–709

Darko, A., Chan, A.P.C., Adabre, M.A., Edwards, D.J., Hosseini, M.R. and Ametwa, E.E. (2020). Artificial intelligence in the AEC industry: Scientometric analysis and visualization of research activities. *Automation in Construction*, 112: 1–19

Datta, M. (2000). Challenges facing the construction industry in developing countries. *Paper Presented at the 2nd International Conference on Construction in Developing Countries*, Gaborone, November, pp. 119–127

Delgado Camacho, D., Clayton, P., O'Brien, W.J., Seepersad, C., Juenger, M., Ferron, R. and Salamone, S. (2018). Applications of additive manufacturing in the construction industry – a forward-looking review. *Automation in Construction*, 89: 11–119

Department of Public Works. (2014). Infrastructure sector plan for expanded public works programme. Available at: www.publicworks.gov.za [accessed 15–8–2018]

Dickinson, H. (2018). The next industrial revolution? The role of public administration in supporting government to oversee 3D printing technologies. *Public Administration Review*, 78(6): 922–925

Dimick, S. (2014). *Adopting Digital Technologies: The Path for SMEs*. The Conference Board of Canada, Ottawa, ON, pp. 1–13

Doller, J. and Hegedorn, B. (2008). *Integrating Urban GIS, CAD and BIM Data by Service-Based Virtual 3D City Models, Urban and Regional Data Management*. UDMS Annual 2007, Taylor and Francis, London, pp. 157–170

Dong, X., Chang, Y., Wang, Y. and Yan, J. (2017). Understanding usage of internet of things (IOT) systems in China: Cognitive experience and affect experience as moderator. *Information Technology & People*, 30(1): 117–138

Eadie, R., Odeyinka, H., Browne, M., McKeown, C. and Yohanis, M. (2013). An analysis of the drivers for adopting building information modelling. *Journal of Information Technology in Construction (ITcon)*, 18(17): 338–352

Eastman, C., Teicholz, P., Sacks, R. and Liston, K. (2011). *BIM Handbook – A Guide to Building Information Modeling for Owners, Managers, Designers, Engineers and Contractors*, 2nd edition. Wiley and Sons, NJ

Fonseca, L.M. (2018). Industry 4.0 and the digital society: Concepts, dimensions and envisioned benefits. *Proceedings of the International Conference on Business Excellence*, 34: 386–397

Fox, S. and Hietanen, J. (2007). Interorganizational use of building information models: Potential for automational, informational and transformational effects. *Construction Management and Economics*, 25(3): 289–296

Gaille, B. (2016). 11 Pros and cons of technology in business. Available at: https://brandon-gaille.com/11-pros-and-cons-of-technology-in-business/ [accessed 15–8–2018]

Gamil, Y.A., Abdullah, M., Abd Rahman, I. and Asad, M.M. (2020). Internet of things in construction industry revolution 4.0: Recent trends and challenges in the Malaysian context. *Journal of Engineering, Design and Technology*, 18(5): 1091–1102

Gheisari, M., Williams, G., Walker, B. and Irizarry, J. (2014). Locating building components in a facility using augmented reality Vs. Paper-based methods: A user-centred experimental comparison. *Proceeding of the International Conference on Computing in Civil and Building Engineering*, Atlanta, June, pp. 850–857

Gilkinson, N., Raju, P., Kiviniemi, A. and Chapman, C. (2015). Building information modelling: The tide is turning. *Structures and Buildings*, 168(SB2): 81–93

Goldin, M. (2014). Chinese company builds houses quickly with 3d printing. Available at: http://mashable.com/2014/04/28/3d-printing-houses-china/ pdf [accessed 15–8–2018]

Golizadeh, H., Hosseini, M.R., Edwards, D.J., Abrishami, S., Taghavi, N. and Banihashemi, S. (2019). Barriers to adoption of RPAs on construction projects: A task – technology fit perspective. *Construction Innovation*, 19(2): 149–169

Gopinath, V., Srija, A. and Sravanthi, C.N. (2019). Re-design of smart homes with digital twins. *Journal of Physics: Conference Series*, 1228(1): 1–10

Gravley, D. (2001). *Risk, Hazard, and Disaster*. University of Canterbury, New Zealand

Gromov, V.A., Voronin, I.M., Gatylo, V.R. and Prokopalo, E.T. (2015). Active cluster replacement algorithm as a tool to assess bifurcation early-warning signs for von Karman equations. *Artificial Intelligence Research*, 6(2): 51–56

Gubbi, J., Buyya, R., Marusic, S. and Palaniswami, M. (2013). Internet of things: A vision, architectural elements, and future directions. *Future Generation Computer Systems*, 29(7): 1645–1660

Guz, A.N. and Rushchitsky, J.J. (2009). Scopus: A system for the evaluation of scientific journals. *International Applied Mechanics*, 45(4): 351–362

Hager, I., Golonka, A. and Putanowicz, R. (2016). 3D printing of buildings and building components as the future of sustainable construction? *Procedia Engineering*, 151: 292–299

Hamdi, O. and Leite, F. (2014). Conflicting side of building information modelling implementation in the construction industry. *Journal of Legal Affairs and Dispute Resolution in Engineering and Construction*, 6(3): 1–8

Haron, A.T., Marshell-Ponting, A. and Aouad, G. (2010). Building information modelling: Literature review on model to determine the level of uptake by organisations. *Proceedings of the 18th CIB World Building Congress 2010*, The Lowry, Salford Quays, 10–13 May, pp. 168–184

Hawkins, D.T. (2001). Bibliometrics of electronic journals in information science. *Information Research*, 7(1): 1. Available at: http://InformationR.net/ir/7-1/paper120.html [accessed 19–12–2019]

Hermann, M., Pentek, T. and Otto, B. (2016). Design principles for industrie 4.0 scenarios. *Proceeding of the 49th Hawaii International Conference on System Sciences (HICSS)*, Kauai, Hawaii, 5–8 January

Heydarian, A., Carneiro, J.P., Gerber, D., Becerik-Gerber, B., Hayes, T. and Wood, W. (2015). Immersive virtual environments versus physical built environments: A benchmarking study for building design and user-built environment explorations. *Automation in Construction*, 54: 116–126

Hirsch, J. (2005). An index to quantify an individual's scientific research output. *Proceedings of the National Academy of Sciences*, 102(46): 16569–16572

Hlahla, P. (2013). *The Use of Information and Communications Technology in the Construction Sector in Gauteng: A Case Study of Khuthaza Affiliated Contractors*. University of the Witwatersrand, Johannesburg

Howard, R., Restrepo, L. and Chang, C. (2017). Addressing individual perceptions: An application of the unified theory of acceptance and use of technology to building information modelling. *International Journal of Project Management*, 35: 107–120

Hudson, V. (2017). The digital future of the infrastructure industry: Innovation 2050. *Balfour Beatty*. Available at: www.balfourbeatty.com/how-we-work/public-policy/innovation-2050-a-digital-future-for-the-infrastructure-industry/ [accessed 19–12–2019]

Institute of Management Development (IMD). (2018). World digital competitiveness ranking. Available at: www.imd.org/wcc/world-competitiveness-center-rankings/world-digital-competitiveness-rankings-2018/ [accessed 2–12–2020]

International Telecommunication Union. (2014). Internet of things global standards initiative. Available at: www.itu.int/en/ITU-T/gsi/iot/Pages/default.aspx#:~:text=IoT%2 DGSI%20aimed%20to%20promote,Things%20on%20a%20global%20scale [accessed 1–12–2020]

Irizarry, J., Gheisari, M., Williams, G. and Walker, B.N. (2013). InfoSPOT: A mobile augmented reality method for accessing building information through a situation awareness approach. *Automation in Construction*, 33: 11–23

Isıkdag, U., Aouad, G., Underwood, J. and Wu, S. (2007). Building information models: A review on storage and exchange mechanisms. *Proceedings of CIB W78*, Maribor, Slovenia.

Ismail, S.A., Bandi, S. and Maaz, Z.N. (2018). An appraisal into the potential application of big data in the construction industry. *International Journal of Built Environment and Sustainability*, 5(2): 145–154

Issa, R. (2002). A survey of e-business implementation in the US construction industry. Rinker school of building construction, University of Florida, Gainesville. *Journal of Information Technology in Construction (ITcon)*, 8: 15–28

Jayasena, H.S. and Weddikkara, C. (2013). Assessing the BIM maturity in a BIM infant industry. *Proceedings of the Second World Construction Symposium: Socio-Economic Sustainability in Construction*, Colombo, Sri Lanka, 14–15 June

Jin, R., Gao, S., Cheshmehzangi, A. and Aboagye-Nimo, E. (2018). A holistic review of off-site construction literature published between 2008 and 2018. *Journal of Cleaner Production*, 20: 1202–1219

Jin, R., Zou, P.X.W., Piroozfar, P., Wood, H., Yang, Y., Yan, L. and Han, Y. (2019). A science mapping approach-based review of construction safety research. *Safety Science*, 113: 285–297

Jin, Y. and Yu, J. (2016). Research on 3D printing in China: A review. *International Journal of Research Studies in Science, Engineering and Technology*, 3(7): 54–60

Johnson Controls. (2019). Applying digital twins to the built environment. Available at: www.johnsoncontrols.com/-/media/jci/insights/2019/bts/jci-661_dv_digital_twin_white_paper_020819_6p_f3.pdf [accessed 19–10–2019]

Jung, W. and Lee, G. (2015). The status of BIM adoption on six continents. *World Academy of Science, Engineering and Technology International Journal of Civil and Environmental Engineering*, 9(5): 512–516

Kagermann, H. (2014). *How Industry 4.0 Will Coin the Economy of the Future? (The Results of the German High-Tech Strategy's and Strategic Initiative Industry 4.0)*. Royal Academy of Engineering, London, February.

Kalavendi, R. (2017). Digitisation and digitalisation: The two-letter difference. Available at: https://medium.com/@ravisierraatlantic/digitization-digitalization-the-2-letter-difference-59b747d42ade [accessed 19–11–2020]

Kamaruddin, S., Mohammed, M.F. and Mahbub, E. (2016). Barriers and impact of mechanisation and automation in construction to achieve better quality products, Malaysia. *Procedia – Social and Behavioural Sciences*, 222: 111–120

Kane, G., Palmer, D., Phillips, A. and Kiron, D. (2015). Is your business ready for a digital future? Available at: https://sloanreview.mit.edu/article/is-your-business-ready-for-a-digital-future/ [accessed 19–11–2020]

Katal, A., Wazid, M. and Goudar, R.H. (2013). Big data: Issues, challenges, tools and good practices. *Proceedings of the Sixth International Conference on Contemporary Computing (IC3)*, Noida, India, 8–10 August, pp. 404–409

Khoshnevis, B. (2003). Toward total automation of on-site construction – An integrated approach based on contour crafting. *Proceedings of the 20th ISARC*, Eindhoven, Holland, pp. 61–66

Kifokeris, D. and Koch, C. (2019). Blockchain in construction – hype, hope, or harm? Advances in ICT in design, construction and management in architecture, engineering, construction and operations (AECO). *Proceedings of the 36th CIB W78 2019 Conference*, Northumbria University, Newcastle, United Kingdom, 18–20 September, pp. 189–198

Kim, M.K., Cheng, J.C.P., Sohn, H. and Chang, C.C. (2015). A framework for dimensional and surface quality assessment of precast concrete elements using BIM and 3D laser scanning. *Automation in Construction*, 49: 225–238

Kim, M.K. and Shin, D. (2015). An acceptance model for smart watches. *Internet Research*, 25(4): 527–541

Kim, Y. and Haas, C.T. (2000). A model for automation of infrastructure maintenance using representational forms. *Automation in Construction*, 10(1): 57–68

Kodeboyina, S.M. and Varghese, K. (2016). Low cost augmented reality framework for construction applications. *Proceedings of the 33rd International Symposium on Automation and Robotics in Construction*, Auburn, Alabama, 18–21 July

Kopsida, M., Vela, P. and Brilakis, I. (2015). A review of automated construction progress monitoring methods. *Proceedings of the 32nd CIB W78 Conference*, Eindhoven, The Netherlands, 27–29 October, pp. 421–431

Kovačić, I., Filzmoser, M., Kiesel, K., Oberwinter, L. and Mahdavi, A. (2015). BIM teaching as support to integrated design practice. *Gradevinar*, 67(6): 537–546

KPMG. (2016). Global construction survey. Building a technology advantage. Available at: https://assets.kpmg.com/content/dam/kpmg/xx/pdf/2016/09/global-construction-survey-2016.pdf [accessed 17-7-2018])

Kumar, B. and Cheng, J.C.P. (2010). Cloud computing and its implications for construction IT. *Computing in Civil and Building Engineering, Proceedings of the International Conference*, 30: 315–341

Kwon, O.S., Park, C.S. and Lim, C.R. (2014). A defect management system for reinforced concrete work utilizing BIM, image-matching and augmented reality. *Automation in Construction*, 46: 74–81

Kypriotaki, K. (2015). From bitcoin to decentralized autonomous corporations – extending the application scope of decentralized peer-to-peer networks and blockchains. *Proceedings of the 17th International Conference on Enterprise Information Systems*, SCITEPRESS – Science and Technology Publications, Setubal, Portugal, pp. 284–290

Labonnote, N., Rønnquist, A., Manum, B. and Rüther, P. (2016). Additive construction: State of the art, challenges and opportunities. *Automation in Construction*, 72(3): 347–366

Lanko, A. (2018). Additive technologies in construction of temporary housing for victims of natural disasters and other emergencies. In Murgul, V. and Popovic, Z. (Eds.), *International Scientific Conference Energy Management of Municipal Transportation Facilities and Transport. Advances in Intelligent Systems and Computing*, 692: 1102–1108

Lawrence, M., Roberts, C. and King, L. (2017). *Managing Automation: Employment, Inequality and Ethics in the Digital Age*. IPPR Commission on Economic Justice Available at: www.ippr.org/files/2018-01/cej-managing-automation-december2017.pdf [accessed 21–12–2020]

Lee, S. and Yu, J. (2016). Comparative study of BIM acceptance between Korea and the United States. *Journal of Construction Engineering and Management*, 142(3): 1–9

Le Roux, D. (2018). Automation and employment: The case of South Africa. *African Journal of Science, Technology Innovation and Development*, 10(4): 507–517

Li, C.Z., Xue, F., Li, X., Hong, J. and Shen, G.Q. (2018). An internet of things-enabled BIM platform for on-site assembly services in prefabricated construction. *Automation in Construction*, 86: 146–161

Li, J.G.D. and Kassem, M. (2019). Blockchain in the built environment and construction industry: A systematic review, conceptual models and practical use cases. *Automation in Construction*, 102: 288–307

Li, X., Yi, W., Chi, H.L., Wang, X. and Chan, A.P. (2018). A critical review of virtual and augmented reality (VR/AR) applications in construction safety. *Automation in Construction*, 86: 150–162.

Lim, S., Buswell, R.A., Le, T.T., Austin, S.A., Gibb, A.G.F. and Thorpe, T. (2012). Developments in construction-scale additive manufacturing processes. *Automation in Construction*, 21: 262–268

Lipscomb, H.J., Dement, J.M., Nolan, J. and Patterson, D. (2006). Nail gun injuries in apprentice carpenters: Risk factors and control measures. *American Journal of Industrial Medicine*, 49: 505–513

Love, P., Irani, Z., Li, H., Cheng, E. and Tse, R. (2001). An empirical analysis of the barriers to implementing e-commerce in small-medium sized construction contractors in the state of Victoria, Australia. *Construction Innovation*, 1(1): 31–41

MacRae, M. (2016). The 3D printed office of the future. Available at: www.asme.org/topics-resources/content/3d-printed-office-the-future [accessed 21–12–2020]

Magrassi, P. and Berg, T. (2002). *A World of Smart Objects: The Role of Auto-Identification Technologies*. Research Report No. R-17–2243. Gartner, Stamford

Mahalingam, A., Yadav, A.K. and Varaprasad, J. (2015). Investigating the role of lean practices in enabling BIM adoption: Evidence from two Indian cases. *Journal of Construction Engineering & Management*, 141(7): 1–11

Mahbub, R. (2008). An Investigation into the Barriers to the Implementation of Automation and Robotics Technologies in the Construction Industry. School of Urban Development. Unpublished Thesis, Faculty of Built Environment and Engineering, Queensland University of Technology, Queensland, Australia

Mandičák, T., Mesároš, P. and Kozlovská, M. (2016). Exploitation of cloud computing in management of construction projects in Slovakia. *Organization, Technology and Management in Construction*, 8: 1456–1463

Mason, J. (2017). Intelligent contracts and the construction industry. *Journal of Legal Affairs and Dispute Resolution in Engineering and Construction*, 9(3): 1943–4162

McDonald, S. (2017). It's time to embrace IoT in construction. *ENR: Engineering News-Record*, L13

Meno, T. (2020). *An Assessment of Risk Associated with Digitalisation in the South African Construction Industry*. Unpublished Masters dissertation submitted to the University of Johannesburg, Johannesburg, South Africa

Meža, S., Turk, Ž. and Dolenc, M. (2014). Component based engineering of a mobile BIM-based augmented reality system. *Automation in Construction*, 42: 1–12

Mohd-Tobi, A.L., Omar, S.A., Yehia, Z., Al-Ojail, S., Hashimi, A. and Orhan, O. (2018). Cost viability of 3D printed house in UK. *IOP Conference Series: Materials Science and Engineering*, 319(1): 1–8

Mosly, I. (2017). Applications and issues of unmanned aerial systems in the construction industry. *International Journal of Construction Engineering & Management*, 6(6): 235–239

Muro, M., Liu, S., Whiton, J. and Kulkarni, S. (2017). *Digitalization and the American Workforce*. Metropolitan Policy Program Report, The Brookings Institution, Washington, DC

Nazir, F.A., Edwards, D.J., Shelbourn, M., Martek, I., Thwala, W.D.D. and El-Gohary, H. (2020). Comparison of modular and traditional UK housing construction: A bibliometric analysis. *Journal of Engineering, Design and Technology*, 19(1): 164–186

Newton, J. (2018). Artificial intelligence in the construction industry. *International Journal of Civil Engineering and Technology*, 9(13): 957–962

Nikkas, A., Poulymenakou, A. and Kriaris, P. (2007). Investigating antecedents and drivers affecting the adoption of collaboration technologies in the construction industry. *Automation in Construction*, 6(5): 632–641

O'Boyle, R. (2017). Block capital: How blockchain could change planning. *The Planner*. Available at: www.theplanner.co.uk/features/block-capital-how-blockchain-could-change-planning [accessed 21–12–2020]

Oke, A.E., Aigbvaboa, C.O. and Mabena, S. (2017). Effects of automation on construction industry performance. Second international conference on mechanics, materials and structural engineering (ICMMSE 2017). *Advances in Engineering Research (AER)*, 102: 370–374

Oke, A.E., Aghimien, D.O., Aigbavboa, C.O. and Koloko, N. (2018). Challenges of digital collaboration in the South African construction industry. *Proceedings of the International Conference on Industrial Engineering and Operations Management*, Bandung, Indonesia, 6–8 March, pp. 2472–2482

Olawumi, T.O., Chan, D.W.M. and Wong, J.K.W. (2017). Evolution in the Intellectual structure of BIM research: A bibliometric analysis. *Journal of Civil Engineering and Management*, 23: 1060–1081

Olugboyega, O. and Aina, O.O. (2016). Analysis of building information modelling usage indices and facilitators in the Nigerian construction industry. *Journal of Logistics, Informatics and Service Science*, 3: 1–36

Pärn, A. and Edwards, D. (2019). Cyber threats confronting the digital built environment: Common data environment vulnerabilities and block chain deterrence. *Engineering, Construction and Architectural Management*, 26(2): 245–266

Parveen, R. (2018). Artificial intelligence in construction industry: Legal issues and regulatory challenges. *International Journal of Civil Engineering and Technology*, 9(13): 957–962

Paul, S.C., van Zijl, G.P.A.G., Tan, M.J. and Gibson, I. (2018). A review of 3D concrete printing systems and materials properties: Current status and future research prospects. *Rapid Prototyping Journal*, 24(4): 784–798

Pauwels, P. and Terkaj, W. (2016). EXPRESS to OWL for construction industry: Towards a recommendable and usable ifcOWL ontology. *Automation in Construction*, 63: 100–133

Penzes, B. (2018). *Blockchain Technology in the Construction Industry: Digital Transformation for High Productivity*. Institute of Civil Engineers Report. ICE, London

Peoples, C., Parr, G., McClean, S., Scotney, B. and Morrow, P. (2013). Performance evaluation of green data center management supporting sustainable growth of the internet of things. *Simulation Modelling Practice and Theory*, 34: 221–242

Pereira, T., Santos, C. and Pires, N. (2002). *The Use of Robots in the Construction Industry*. IAHS World Congress on Housing, University of Coimbra – Faculty of Sciences and Technology Department of Civil Engineering – Constructions Laboratory Portugal, Portugal, pp. 1–10

Perrot, A., Rangeard, D. and Pierre, A. (2016). Structural built-up of cement-based materials used for 3D-printing extrusion techniques. *Materials and Structures*, 49(4): 1213–1220

Pîrjan, A. and Petroşanu, D.M. (2013). The impact of 3D printing technology on the society and economy. *Journal of Information Systems & Operations Management*, 7(2): 163–173

Porter, M.E. and Heppelmann, J.E. (2014). How smart, connected products are transforming competition. *Harvard Business Review*, 92(11): 11–64

PwC. (2016). *Highlighting Trends in the South African Construction Industry. SA Construction*, 4th edition, pp. 1–46. Available at: www.pwc.co.za/construction [accessed 17–7–2018]

Qian, X. and Papadonikolaki, E. (2019). The influence of the blockchain technology on trust in construction supply chain management. Advances in ICT in design, construction and management in architecture, engineering, construction and operations (AECO). *Proceedings of the 36th CIB W78 2019 Conference*, Northumbria University, Newcastle, United Kingdom, 18–20 September, pp. 179–188

Raajan, N.R., Suganya, S., Hermanand, R., Sarada, N.S. and Ramanan, S.V. (2012). Augmented reality for 3D construction. *Procedia Engineering*, 38: 66–72

Rad, B.B. and Ahmada, H.A. (2017). Internet of things: Trends, opportunities, and challenges. *International Journal of Computer Science and Network Security*, 17(7): 89–95

Rawai, N.M., Fathi, M.S., Abedi, M. and Rambat, S. (2013). Cloud computing for green construction management. *3rd International Conference on Intelligent System Design and Engineering Applications*, Hong Kong, China, 16–18 January, pp. 432–435

Reffat, R. (2008). Digital architecture and reforming the build environment. *Journal of Architecture and Planning Research*, 25(2): 118–129

Rubin, M. (2006). *Droid Maker: George Lucas and the Digital Revolution*. Triad Publications, Gainesville

Rüßmann, M., Lorenz, M., Gerbert, P. and Waldner, M. (2015). *Industry 4.0: The Future of Productivity and Growth in Manufacturing Industries*. Boston Consulting Group Report, Boston

Ruggiero, A., Salvo, S. and St. Laurent, C. (2016). *Robotics in Construction*. Worcester Polytechnic Institute, Worcester, MA

Sacks, R. and Barak, R. (2010). Teaching building information modelling as an integral part of freshman year civil engineering education. *Journal of Professional Issues in Engineering Education and Practice*, 136(1): 30–38

Sage. (2012). Cloud security with sage construction anywhere. SAGE Construction and Real Estate. Available at: www.teamtag.net/wp-content/uploads/2015/07/Sage-CRE-Whitepaper-Cloud-Computing-and-the-Construction-Industry.pdf [accessed 19–11–2019]

Saka, A.B. and Chan, D.W.M. (2019). A scientometric review and metasynthesis of building information modelling (BIM) research in Africa. *Buildings*, 9(85): 1–21

Saka, A.B., Chan, D.W.M. and Siu, F.M.F. (2020). Drivers of sustainable adoption of building information modelling (BIM) in the Nigerian construction small and medium-sized enterprises (SMEs). *Sustainability*, 12: 1–23

Sakin, M. and Kiroglu, Y.C. (2017). 3D printing of buildings: Construction of the sustainable houses of the future by BIM. *Energy Procedia*, 134: 702–711

Salento, A. (2017). Digitalisation and the regulation of work: Theoretical issues and normative challenges. *Artificial Intelligence and Society Journal*, 33: 369–378

San, K.M., Choy, C.F. and Fung, W.P. (2019). The potentials and impacts of blockchain technology in construction industry: A literature review. *IOP Conference Series: Materials Science and Engineering*, 495(1): 1–10

Sardroud, J.M., Mehdizadehtavasani, M., Khorramabadi, A. and Ranjbardar, A. (2018). Barriers analysis to effective implementation of BIM in the construction industry. *35th International Symposium on Automation and Robotics in Construction (ISARC)*, Berlin, Germany, 20–25 July

Schia, M.H., Trollsås, B.C., Fyhn, H. and Lædre, O. (2019). The introduction of AI in the construction industry and its impact on human behavior. In Pasquire, C. and Hamzeh, F.R. (Eds.), *Proceedings 27th Annual Conference of the International.* Group for Lean Construction, IGLC, Dublin, Ireland

SCImago Journal Rank. (2020). SCImago journal and country rank. Available at: www.scimagojr.com/journalsearch.php?q=16273andtip=sidandclean=0 [accessed 14–6–2020]

Shin, D. and Park, Y.J. (2017). Understanding the internet of things ecosystem: Multi-level analysis of users, society, and ecology. *Digital Policy, Regulation and Governance,* 19(1): 77–100

Shrivastava, R. and Mahajan, P. (2016). Artificial intelligence research in India: A scientometric analysis. *Science & Technology Libraries,* 35(2): 136–151

Siebert, S. and Teizer, J. (2014). Mobile 3D mapping for surveying earthwork projects using an Unmanned Aerial Vehicle (UAV) system. *Automation in Construction,* 41: 1–14

Smallwood, J. J. (2004). Optimum cost: The role of health and safety. In Verster, J. J. P. (Ed.), *International Cost Engineering Council 4th World Congress.* Cape Town, 17–21 April.

Smith, P. (2014). BIM implementation – global strategies. *Procedia Engineering,* 85: 482–492

Stamford. (2015). *Gartner Says 6.4 Billion Connected "Things" Will Be in Use in 2016, Up 30 Percent from 2015.* Gartner, Inc. Available at: www.gartner.com/newsroom/id/3165317 [accessed 12–11–2019]

Stephenson, P. and Blaza, S. (2001). Implementing technological change in construction organisations. *Proceeding of the CIB-W78 International Conference,* Mpumalanga, South Africa, 30 May–1 June, pp. 1–12

Strukova, Z. and Liska, M. (2012). *Application of Automation and Robotics in Construction Work Execution.* Technical University of Košice, Faculty of Civil Engineering, Slovakia

Succar, B. (2009). Building information modelling framework: A research and delivery foundation for industry stakeholders. *Automation in Construction,* 18(3): 357–375

Sung, T.K. (2018). Industry 4.0: A Korea perspective. *Technological Forecasting and Social Change,* 132(C): 40–45

Tambi, A.S., Kolhe, A.R. and Saharkar, U. (2014). Remedies over the obstacles in implementing automation in indian infrastructure projects, *International Journal of Engineering and Technology (IJRET),* 3(5): 606–608

Tapscott, D. (2016). *Blockchain Revolution: How the Technology Behind Bitcoin is Changing Money, Business, and the World.* Penguin, New York

Taylor, M. and Smith, W. (2000). Analysis of risk in construction automation investment. *17th International Symposium on Automation and Robotics in Construction,* Taipei, Taiwan, pp. 1–6

Taylor, M., Wamuziri, S. and Smith, I. (2003). Automated construction in Japan. *Proceedings of the Institution of Civil Engineers-Civil Engineering,* 156(1): 34–41

Turk, Ž. and Klinc, R. (2017). Potentials of blockchain technology for construction management. *Procedia Engineering,* 196: 638–645

Ullah, K., Lill, I. and Witt, E. (2019). An overview of BIM adoption in the construction industry: Benefits and barriers. *Proceedings of the 10th Nordic Conference on Construction Economics and Organization,* Estonia, May, pp. 297–303

Urie, M. (2019). The internet of things in construction. Available at: https://marketintel.gardiner.com/uploads/1901_IoT-in-Construction.pdf [accessed 15–7–2019]

Vaduva-Sahhanoglu, A., Calbureanu-Popescu, M.X. and Smid, S. (2016). Automated and robotic construction-a solution for the social challenges of the construction sector. *Revista de stiinte politice,* 50: 1–11

Vähä, P., Heikkilä, T. and Kilpeläinen, P. (2013). Extending automation of building construction – Survey on potential sensor technologies and robotic applications. *Automation in Construction*, 36: 168–178

Van Eck, N.J. and Waltman, L. (2014). Visualizing bibliometric networks. In Ding, Y., Rousseau, R. and Wolfram, D. (Eds.), *Measuring Scholarly Impact: Methods and Practice*. Springer, Switzerland, pp. 285–320

Vuksic, V.B., Invansic, L. and Vugec, D.S. (2018). A preliminary literature review on digital transformation case studies. *International Journal of Computer and Information Engineering*, 12(9): 737–742

Wang, M., Zhang, J., Jiao, S. and Zhang, T. (2019). Evaluating the impact of citations of articles based on knowledge flow patterns hidden in the citations. *PLoS One*, 14(11): 1–19

Wang, P., Wu, P., Wang, J., Chi, H.L. and Wang, X. (2018). A critical review of the use of virtual reality in construction engineering education and training. *International Journal of Environmental Research and Public Health*, 15(6): 1–18

Watson, A. (2011). Digital Buildings – Challenges and opportunities. *Advances Engineering Informatics Journal*, 25(4): 573–581

Weber, R.H. (2010). Internet of things: New security and privacy challenges. *Computer Law and Security Review*, 26(1): 23–30

Webster, J. and Watson, R. (2002). The past to prepare for the future: Writing a literature. *MIS Quarterly*, 26(2): 13–23

Winfield, M. and Rock, S. (2018). *The Winfield Rock Report: Overcoming the Legal and Contractual Barriers of BIM*. BIM Alliance. Available at: www.maber.co.uk/app/uploads/2018/03/The-Winfield-Rock-Report.pdf [accessed 11–12–2020]

Won, J., Lee, G., Dossick, C. and Messner, J. (2013). Where to focus for successful adoption of building information modelling within organisation. *Journal of Construction Engineering and Management*, 139(11): 1–10

Wong, K. and Fan, Q. (2013). Building information modelling (BIM) for sustainable building design. *Facilities*, 31(3/4): 138–157

Woodhead, R., Stephenson, P. and Morrey, D. (2018). Digital construction: From point solutions to IoT ecosystem. *Automation in Construction*, 93: 35–46

Wu, P., Wang, J. and Wang, X. (2016). A critical review of the use of 3-D printing in the construction industry. *Automation in Construction*, 68: 21–31

Yaghoubi, S., Kazemi, M. and Sakhaifar, M. (2012). ICT technologies, robotic and automation in construction. *International Journal of Basic and Applied Science*, 12(4): 112–116

Yahya, M.Y., Yin, L.H., Yassin, A.B., Omar, R., Robin, R.O. and Kasim, N. (2019). The challenges of the implementation of construction robotics technologies in the construction. *MATEC Web of Conferences*, 266: 1–5

Yan, H. and Damian, P. (2008). Benefits and barriers of building information modelling. *12th International Conference on Computing in Civil and Building Engineering*, Beijing, China

Yin, X., Liu, H., Chen, Y. and Al-Hussein, M. (2019). Building information modelling for off-site construction: Review and future directions. *Automation in Construction*, 101: 72–91

Young Jr., N.W., Jones, S.A. and Bernstein, H.M. (2008). *Building Information Modelling (BIM): Transforming Design and Construction to Achieve Greater Industry Productivity*. McGraw Hill Construction, New York.

Yu, W.D. and Lin, H.W. (2006). A VaFALCON neuro-fuzzy system for mining of incomplete construction databases. *Automation in Construction*, 15(1): 20–32

Zaychenko, I., Smirnova, A. and Borremans, A. (2018). Digital transformation: The case of the application of drones in construction. *MATEC Web of Conferences*, 193: 1–7

Zeng, Y., Wang, L., Deng, X., Cao, X. and Khundker, N. (2012). Secure collaboration in global design and supply chain environment: Problem analysis and literature review. *Computers in Industry*, 63: 545–556

Zhang, C. (2013). A novel triangle mapping technique to study the h-index based citation distribution. *Scientific Reports*, 3(1): 10–23

Zhao, J., Zheng, X., Dong, R. and Shao, G. (2013). The planning, construction, and management toward sustainable cities in China needs the environmental internet of things. *International Journal of Sustainable Development and World Ecology*, 2(2): 1–4

Zhao, X. (2017). A scientometric review of global BIM research, Analysis and visualisation. *Automation in Construction*, 80: 37–47

Zheng, X., Le, Y., Chan, A.P.C., Hu, Y. and Li, Y. (2016). Review of the application of social network analysis (SNA) in construction project management research. *International Journal of Project Management*, 34(7): 1214–1225

Zhong, R.Y., Neman, S.T., Huag, G.Q. and Lan, S. (2016). Big data for supply chain management in the service and manufacturing sectors: Challenges, opportunities, and future perspectives. *Computers and Industrial Engineering*, 101: 572–591

Zhou, W., Whyte, J. and Sacks, R. (2012). Construction safety and digital design: A review. *Automation in Construction*, 22: 102–111

Zhu, W.S., Zhang, Q.B., Zhu, H.H., Li, Y., Yin, J.-H., Li, S.C., Sun, L.F. and Zhang, L. (2010). Large-scale geomechanical model testing of an underground cavern group in a true three-dimensional (3-D) stress state. *Canadian Geotechnical Journal*, 47(9): 935–946

Part III

Theoretical perspective of digitalisation capability maturity

Part III of this book focuses on the theoretical dimension of digitalisation capability maturity. This part consists of three chapters. Chapter 5 sheds light on the need for a capability maturity model, while Chapter 6 evaluates relevant maturity models. Chapter 7 explores the dynamic capability theory which underpins the proposed construction digitalisation capability maturity model.

5 Why a capability maturity model?

Introduction

This chapter gives the description of a capability maturity model. Drawing from the original capability maturity model developed at the Software Engineering Institute at Carnegie Mellon University, the chapter explains what this type of model entails. Furthermore, the chapter gives direction on what a maturity model should contain by exploring the different phases in developing a maturity model, the maturity model's characteristics, and the different maturity levels required in an ideal capability maturity model.

What is a maturity model?

By description, maturity models show the evolution of an organisation's capabilities based on certain defined stages along a projected, anticipated, or logical path (Röglinger *et al.*, 2012). This submission was based on the assumption that there exists a predictable pattern in organisational evolution and change. Prananto *et al.* (2003) submitted that maturity models are also termed stages-of-growth models, stage models, or stage theories. Röglinger *et al.* (2012) gave the early examples of maturity models to include the hierarchy of human needs by Maslow in 1954, economic growth by Kuznets in 1965, and IT's progression in organisations by Nolan in 1973 and 1979. Philip Crosby, while analysing quality principles, gave five evolutionary stages in adopting quality practices, and these evolutionary stages birth the maturity concept. This concept was later modified into the Capability Maturity Model (CMM) for software processes developed at the Software Engineering Institute (SEI) at Carnegie Mellon University as a means of improvement suggested for software organisations that wish to improve their software process capability (Paulk *et al.*, 1991). The CMM proposed five levels of increasing maturity that these organisations need to attain to improve their software process. The model indicates that an organisation's maturity rises with each level, and each maturity level provides a layer in the foundation for continuous process improvement. Paulk *et al.* (1993) further stated that each of the levels contains a set of goals that, when satisfied, stabilises an important component in the process, thus, increasing the process capability of the organisation.

Maturity model has also been described as a sequence of stages used to assess situations and guide potential improvements within an organisation or industry (Macchi & Fumagalli, 2013; Röglinger *et al.*, 2012). It is a self-assessment tool designed to establish the maturity level of an organisation and at the same time identify the possible potential for improvement or innovation. Therefore, with maturity models, the status quo within an organisation can be measured and, in the process, improvement measures that best suits the organisation can be developed (Essman, 2009). It has also been observed that in more recent times, maturity models are used as a tool for self-assessment, benchmarking, change management, and even organisational learning (Dyk & Schutte, 2012; Kirrane, 2009). The importance of this type of modelling tool to the digitalisation of organisations was emphasised in the report of Capgemini Consulting (2014), which stated that for digitalisation within the manufacturing industry, companies should undertake a detailed digital maturity assessment to have a transparent view on their current level of digital readiness. This can be seen as a self-assessment process for these companies regarding their digitalisation level and readiness. Similarly, if construction companies are to achieve headway to increase their service delivery capabilities, then a digital maturity model that will serve as a road map towards achieving this quest is necessary.

Due to the simple nature of most maturity models, some studies have described maturity models as stepwise models that tend to oversimplify the actual reality within organisations and are, in most cases, deficient of empirical backings. Therefore, it has been suggested that in developing a maturity model, there should be a balance between the complex reality that exists within organisations and the simplicity of the developed model. This is because an oversimplified model may not adequately reflect the complexities of the actual situation being assessed, while a complex model may limit interest or create confusion. The result of any of these two scenarios will be the wrong application of the developed model and misleading outcomes (De Bruin *et al.*, 2005; McCormack *et al.*, 2009). Also, it has been mentioned that maturity models should be configurable, since internal and external characteristics may affect a maturity model's applicability in its standard version (Mettler & Rohner, 2009). Röglinger *et al.* (2012) noted that maturity models have come under fire in areas of the high number of similar maturity models being developed, dissatisfaction documentation, poor adoption of the CMM blueprint known as the source from which other maturity models emanate, and missing economic foundation. Despite these criticisms, several maturity models are still being developed to provide effective methods of service delivery in diverse fields and industries with the implementation of suggestions that have been made for effective maturity models (Becker *et al.*, 2009; De Bruin *et al.*, 2005; Maier *et al.*, 2012).

Since the development of the CMM, several maturity models have been developed based on its principles. A more significant improvement of the CMM is the Capability Maturity Model Integration (CMMI) which incorporates three legacies of the CMMs into one (Ahern *et al.*, 2004; De Bruin *et al.*, 2005). Others include the Change Management Maturity Model (CM3) developed by Sun *et al.* (2009) to handle change in construction projects; the Change Management

Capability Maturity Level (CMCML) developed by Arowosegbe and Sarajul (2015) also geared towards addressing change issues in construction by contracting firms; the Analysis Capability Maturity Model (ACMM), which was developed for the US-American National Reconnaissance Office with the sole purpose of evaluating the processes of organisations that carry out state-commissioned projects (Covey & Hixon, 2005); Project Management Maturity Model (PM3); Organisational Project Management Maturity Model (OPM3); the Standardised Process Improvement for Construction Enterprises model (SPICE) reported by Sarshar *et al.* (1999); Business Process Management Maturity (BPMM) developed by Rosemann *et al.* (2006). Vivares *et al.* (2018), in a systematic literature review, discovered articles on 42 maturity models. De Bruin *et al.* (2005) stated that about 150 maturity models had been developed over time in different domains. This is a pointer to the importance of maturity models to organisation activities and process development. Interestingly, while these maturity models exist, there is a lack of recorded information on a maturity model targeted at digitalisation issues within the construction industry. This non-existence of maturity model to assess digitalisation within the construction industry shows a knowledge gap that requires further investigation.

Developing a maturity model

The scientific backing of a maturity model is rooted in design science, which tries to extend the boundaries of human and organisational capabilities through the creation of innovative artefacts (Boström & Celik, 2017). "Design" in this context is a process of solving a problem, and an "artefact" is an outcome of such a process (Hevner *et al.*, 2004). This artefact can be in the form of a model, method, or construct that has been adopted to solve a particular problem. Becker *et al.* (2009), therefore described a maturity model as an artefact that can tell an organisation's current capabilities position and help proffer a possible line of actions for change and improvement.

In developing a maturity model, Becker *et al.* (2009) proposed an eight-requirement approach. These requirements follow the design science research guidelines earlier postulated by Hevner *et al.* (2004). The first thing to consider is whether there is a need for a new model to solve the specific problem or just the need to improve already existing ones. To determine this, there must be a comparison with existing models. The review of existing maturity models in digitalisation shows that only a few digital maturity models exist in the IT, manufacturing, and business domains. No visible evidence can be seen as regards a maturity model in digitalisation of the construction industry. Thus, following the concept of the few available models, and contextualising it in the construction industry to fit in with the activities of the industry is crucial in the development of a digitalisation maturity model for the construction industry (Becker *et al.*, 2009).

After comparison, an iterative and step-by-step process must be adopted in developing the model. This step-by-step process would help refine, evaluate, and enhance the necessary components within the model to be designed. The third requirement is crucial as it involves evaluating the principles, premises, usefulness, effectiveness, and even the quality of the maturity model. Hevner *et al.* (2004) noted that since an artefact's design may follow different developmental stages, the use of an appropriate scientific methodology in its evaluation is necessary. Therefore, it was stated that the usefulness, quality, and efficiency of a designed model or method must be thoroughly demonstrated through the use of well-executed evaluation methods. The fourth requirement is a multi-methodological procedure. In developing a maturity model, diverse research methods are adopted, and these research methods need to be reliable and finely attuned to the problem the model is to solve (Boström & Celik, 2017). Through this reliable and finely adapted research methodology, the developed model can contribute significantly to the identified problem.

The fifth requirement is the proper identification of problem relevance. Hevner *et al.* (2004) stressed that not only should a problem-solving artefact be innovative, it should also be relevant to researchers and practitioners within the field of such problem. Therefore, Becker *et al.* (2009) noted that this relevance could be established through different scientific methods such as interviewing potential users of the developed artefact. Aside from determining the significance of the problem, the prospective domain wherein the developed maturity model will be used, as well and its usage and the outcomes must be clearly defined before designing of the model. This process is known as the "problem definition" process, and it is the sixth requirement. As stated earlier in the research problem in Chapter 1, the construction industry in most developing countries is slow to adopt digital technologies; hence it has failed to enjoy its benefits. This failure is in most cases due to lack of knowledge on how to go about attaining digital transformation. Thus, if the construction industry in these countries is to be digitalised and better services provided through digitalisation, a road map for achieving digitalisation is needed. Herein lies the relevance of developing a digitalisation capability maturity model that will tell the capability of construction organisations in attaining digital transformation and serve as a tool to guide these organisations to move from their present digitalisation level to a more established stage. It will also create a benchmark for them to evaluate their current capabilities and improve.

The seventh requirement is the targeted presentation of results, which involves presenting the developed maturity model in line with the conditions of its application and the needs of its users. Boström and Celik (2017) see this requirement as geared towards presenting the maturity model to the identified users and making it accessible in a planned or predefined way. The last requirement is the scientific documentation, which requires the documenting of every step, process, and procedure adopted in designing the maturity model and the parties involved and the result obtained. This aspect of documentation was deemed crucial following Hevner *et al.'s* (2004) submission that researchers who use developed models in most cases make specific demands for both the results and the research process. The same can be provided through this documentation, and future development can be made on such developed models for more efficient outcome in the future.

Phases in developing a maturity model

De Bruin *et al.* (2005) gave the different phases involved in a generic framework/ model as scope, design, populate, test, deploy, and maintain. These six phases are sequential and are evident in the development and use of a model. It was noted that the decisions taken at the scoping stage, which is the initial stage where the scope of the model is determined, will affect the method of populating the model and the method of testing the developed model. Aside from being sequential, each phase's progression may be iterative as the result of a step might mean the repetition or revisiting of previous steps to confirm the obtained result.

Phase 1: The scoping phase

At the scoping phase, the boundaries of the model's application are set, and the focus of the model in terms of its domain is determined. Focusing the model in a specific domain will help distinguish the model from other existing general maturity models. In this book, the focus of the developed maturity model is the digitalisation of organisations in the construction industry, which is deficient in a maturity model on digitalisation. Also, the development stakeholders are established at this phase. These stakeholders will be involved in the development of the model. The determining of the stakeholders to be used in the development of the model is in most cases influenced by the model's purpose (De Bruin *et al.*, 2005).

Phase 2: The design phase

The design phase incorporates the needs of the intended audience and how these needs will be met. This phase follows the already discussed eight requirements needed to design a maturity model as given by Becker *et al.* (2009). At this phase, the audience (internal or external) is determined, and the method of application, either through self-assessment, third party assisted, or using certified practitioners, is also established. The drivers of applying the developed model, the respondents of the application (management, staff, business partners, etc.), and the mode of the application itself (single or multiple entities, or single or multiple regions) are also determined in this phase.

Phase 3: The populate phase

At the populate phase, the content of the model is decided. Based on existing models and review of theoretical works, the model's major components and sub-components are determined. It was established that the identification of components related to the specific domain for which the model is being created is critical especially for complex domains (construction industry for example) as this will help give a better depth in understanding the maturity of such domain, without which the identification of specific strategies for improving such domain will be difficult (De Bruin *et al.*, 2005). Rosemann and De Bruin (2004) have earlier stated that one of the best ways to gain insight into a domain component is to evaluate critical success factors and barriers in that specific domain. However, De Bruin *et al.* (2005) suggested using an exploratory research method like Delphi, nominal group technique, case study interviews, and focus group discussion in getting more related constructs, especially in complex domains.

Phase 4: The testing phase

After populating, the next phase deals with testing the populated model's constructs and the model instrument. This is done to check the relevance of the model and its rigour. Aside from testing the model's construct, the validity and reliability of any assessment instrument used are deemed crucial. Validity such as the convergence and divergence of the instruments can be done using factor analysis (Freeze & Kulkarni, 2005). Also, a pilot survey can be conducted to ensure the ease of the instrument. The pilot group individuals can be selected from the larger population of the study identified during the scoping phase.

Phase 5: The deployment phase

When the relevance, rigour, validity, and reliability of the populated model have been tested, the model can then be deployed by making it available to the identified audience to determine its generalisability. Deployment includes applying the model in organisations initially identified from the onset of the design. However, where a model has been developed and tested using an involved organisation or stakeholder's resources, the deployment is mostly done within such an organisation or using a stakeholder. Through this deployment, the critical issues relating to the model's ability to be generalised and accepted can be determined (De Bruin *et al.*, 2005).

Phase 6: The maintaining phase

It is not enough to design and deploy a maturity model, maintaining it throughout its growth and use is crucial. To successfully establish a maturity model's generalisability, provisions must be made to handle the high volume of the model's application. Tracking the evolution of the model, which can be a result of broadening and deepening domain knowledge and model understanding, is necessary. If the developed model is prescriptive in nature (a model that prescribes guidelines towards attaining maturity), then there should be resources put in place to track interferences during application. This will further support the model's standardisation and global acceptance (De Bruin *et al.*, 2005).

Characteristics of a maturity model

Carnegie Mellon University (2005) described a maturity model as one that provides a place to start for organisations, provides the benefit of a community's prior experiences, provides a common language and a shared vision, provides a framework for prioritising actions, and provides a way to define what improvement means to an organisation. According to De Bruin *et al.* (2005), a maturity model can be descriptive, prescriptive, or comparative.

Descriptive model is designed to assess the "as-is" situation of an organisation or a process. This type of model is seen as a "one-off" application which creates no room for maturity improvement or creation of avenue for comparing organisation capability with performance.

The prescriptive model focuses on the domain relationships to performance and clearly shows the approach needed to improve maturity that will positively affect the business value. This type of model gives a clear direction to find solution to problems as easy as possible.

Comparative model allows benchmarking across organisations, industries, or regions. This type of model can compare similar practices across organisations to benchmark the maturity within different industries.

Although developed maturity models can exist in these three different forms, an ideal model will go through the different phases throughout its life cycle. A model can start as being descriptive to gain a better understanding of the situation within the domain for which it is designed. The model then evolves into a prescriptive model by understanding the current situation to prescribing informed guidelines for repeatable improvements. With the adequate application of the model within a wide range of organisations to harness adequate data for comparison, the model can become a comparative model that organisations can use as a benchmarking tool (De Bruin *et al.*, 2005).

Other studies have identified some key design features of a maturity model, and these are having relevance to the organisational structure, culture, and working practices; being aligned to the strategic and tactical priorities of the organisation; independent of the technology and vendor; being technical and non-technical IT infrastructure capabilities; easy to maintain and extend based on the changes in an organisation's needs and priorities; intuitive and straightforward to use; can be used for internal and external benchmarking, self-assessment, change management, and organisational learning (Kirrane, 2009; Savidas 2009; Van de Waeering & Batenburg, 2009). De Bruin *et al.* (2005) noted that a maturity model should be structured hierarchically into multiple layers in terms of its structure. Fraser *et al.* (2002) also submitted that the major components of a maturity model are levels, descriptors, descriptions for each group, capability areas, activities for each capability area, and a description of each activity performed at a certain maturity level.

Maturity levels

In most maturity models, a common design principle is the representation of maturity as several cumulative stages with higher stages build on the requirements of lower stages. Wendler (2012) stated that "*these stages are sequential in nature and they represent a hierarchical progression and are closely connected to organisational structures and activities, and stages which measure the completeness of the analysed objects via different sets of (multi-dimensional) criteria".* Using a Likert scale, five represents high maturity and one represents a low maturity level in numbering. This approach was made popular with the CMM, and its practical acceptance has become obvious in recent times (De Bruin *et al.*, 2005). However, while the five-maturity level system is common, evidence of other numbers ranging from three to seven in terms of maturity levels abound (Vivares *et al.*, 2018). Although the number of stages may vary in different models, the crucial aspect of the maturity model is that the final stages are distinct and well defined (De Bruin *et al.*, 2005).

The CMM defined its five levels of maturity as follows:

Level one is described as the "Initial level" where organisations that fall at this level do not have a stable and friendly environment for software development and maintenance. Organisations lack the right management practices, and the benefits of being derived from acceptable software engineering practices are not being enjoyed due to ineffective planning.

Level two is the "repeatable level" where policies for managing a software project and procedures to implement those policies are established. This level is repeatable because the experience harnessed from similar projects handled over time is used to plan and manage new projects.

Level three is described as "defined level". This is the level whereby the standard process for developing and maintaining software across the organisation is documented. This documentation includes both the engineering and management processes of the software. These processes are then combined into a single entity.

Level four is seen as the "managed level" where organisations set quantitative quality goals for software products and processes. At this level, productivity is measured as part of the organisation's measurement programme. Also, software processes are instrumented with well-defined and consistent measurements.

Level five is described as "optimising level". At this level, the entire organisation is focused on continuous process improvement. The means to identify weaknesses and strengthen the process proactively, with the goal of preventing the occurrence of defects is available within organisations at this level (Paulk *et al.*, 1993).

Since no digitalisation maturity model exists within the construction industry, whose maturity level can be adapted, drawing from the origin of maturity models was deemed reasonable in developing one. Thus, Chapter 8 of this book adopted the CMMI's five-maturity level, which is a modification of the CMM. The maturity levels are described as 1- Initiating level, 2 – repeatable level, 3 – defined level, 4 – managed level, and 5 – optimising level (Carnegie Mellon SEI, 2005;

Paulk *et al.*, 1993). These maturity levels are further described under the CMMI in the evaluation of relevant maturity models.

Summary

The chapter explored the concept of a maturity model to give a firm theoretical background of the intended DCMM that the book hopes to develop. The chapter concluded that a capability maturity model is most suitable for the digitalisation of construction organisations as a maturity model will allow construction organisations to measure their existing capability, and in the process, create solutions that best suit the organisation. The chapter revealed that an ideal capability maturity model should have clearly defined capability process areas and maturity levels, with each level building on the preceding one. This maturity model can be designed to be descriptive to better understand the situation within the domain for which it is intended, or prescriptive to understand the current situation and prescribe informed guidelines for repeatable improvements, or comparative by creating a benchmark for organisations to follow. The next chapter of the book discusses the different related maturity models which will serve as the bases for developing the proposed DCMM.

References

Ahern, D.M., Clouse, A. and Turner, R. (2004). CMMI *Distilled: A Practical Introduction to Integrated Process Improvement*, 2nd edition. Addison-Wesley, Boston and London

Arowosegbe, A.A. and Sarajul, F.A. (2015). Towards change management capability assessment model for contractors in building project. *Middle-East Journal of Scientific Research*, 23(7): 1327–1333

Becker, J., Knackstedt, R. and Pöppelbuß, J. (2009). Developing maturity models for IT management. *Business and Information Systems Engineering*, 1(3): 213–222

Boström, E. and Celik, O.C. (2017). *Towards a Maturity Model for Digital Strategizing – A Qualitative Study of How an Organisation Can Analyse and Assess Their Digital Business Strategy*. IT Management Master Thesis submitted to the Department of informatics, UMEA Universitet, Sweden

Capgemini Consulting. (2014). Digitising manufacturing: Ready, set, go! Available at: www.de.capgemini-consulting.com/resource-fileaccess/resource/pdf/digitizing-manufac turing_0.pdf [accessed 19–11–2020]

Carnegie Mellon SEI. (2005). *Capability Maturity Model® Integration (CMMI) Overview*. Carnegie Mellon University, Pittsburgh, PA

Covey, R.W. and Hixon, D.J. (2005). The creation and use of an analysis capability maturity model (ACMM). Available at: http://stinet.dtic.mil/cgibin/GetTRDoc?AD=ADA4364 26andLocation=U2anddoc=GetTRDoc.pdf

De Bruin, T., Freeze, R., Kaulkarni, U. and Rosemann, M. (2005). Understanding the main phases of developing a maturity assessment model. In Campbell, B., Underwood, J. and Bunker, D. (Eds.), *Australasian Conference on Information Systems (ACIS)*, Australia, New South Wales, Sydney, 30 November–2 December

Dyk, L.V. and Schutte, C.S.L. (2012). Development of a maturity model for telemedicine. *South African Journal of Industrial Engineering*, 23(2): 61–72

Essman, H. (2009). *Toward Innovation Capability Maturity*. PhD thesis, Stellenbosch University, Stellenbosch, South Africa

Fraser, P., Moultrie, J. and Gregory, M. (2002). The use of maturity models/grids as a tool in assessing product development capability. *Proceeding of the IEEE International Engineering Management Conference*, Cambridge, 18–20 August

Freeze, R.D. and Kulkarni, U. (2005). Knowledge management capability assessment: Validating a knowledge assets measurement instrument. *Proceedings of the Hawaii International Conference on System Sciences, HICCS-38*, Kauai, Hawaii

Hevner, A.R., March, S.T., Park, J. and Ram, S. (2004). Design science in information systems research. *MIS Quarterly*, 28(1): 75–105

Kirrane, J. (2009). *A Maturity Model for Continuous Quality Improvement of a Clinical Information System Used in Critical Care Medicine*. College of Business, Public Policy and Law, National University of Ireland, Galway, Ireland

Macchi, M. and Fumagalli, L. (2013). A maintenance maturity assessment method for the manufacturing industry. *Journal of Quality in Maintenance Engineering*, 19(3): 295–315

Maier, A.M., Moultrie, J. and Clarkson, P.J. (2012). Assessing organisational capabilities: Reviewing and guiding the development of maturity grids. *IEEE Transactions on Engineering Management*, 59(1): 138–159

McCormack, K., Willems, J., van den Bergh, J., Deschoolmeester, D., Willaert, P., Stemberger, M.I., Skrinjar, R., Trkman, P., Ladeira, M.B., Valadares de Oliveira, M.P., Vuksic, V.B. and Vlahovic, N. (2009). A global investigation of key turning points in business process maturity. *Business Process Management Journal*, 15(5): 792–815

Mettler, T. and Rohner, P. (2009). Situational maturity models as instrumental artifacts for organisational design. *Proceedings of the 4th International Conference on Design Science Research in Information Systems and Technology*, Philadelphia, PA, May, pp. 1–9

Paulk, M., Curtis, B. and Chrissis, M.B. (1991). Capability maturity model for software, engineering Institute, CMU/SEI-91-TR-24, ADA existing buildings, 240603

Paulk, M., Curtis, B., Chrissis, M.B. and Weber, C. (1993). Capability maturity model for software, version 1.1. Available at: http://wwwsei.cmu.edu/pub/documents/93.reports/pdf/tR24.93.pdf

Prananto, A., McKay, J. and Marshall, P. (2003). A study of the progression of e-business maturity in Australian SMEs: Some evidence of the applicability of the stages of growth for e-business model. A paper presented at Pasific Asia Conference on Information Systems (PACIS), Adelaide

Röglinger, M., Pöppelbuß, J. and Becker, J. (2012). Maturity models in business process management. *Business Process Management Journal*, 18(2): 328–346

Rosemann, M. and De Bruin, T. (2004). Application of a Holistic Model for Determining BPM Maturity. *Proceedings of the AIM Pre-ICIS Workshop on Process Management and Information Systems*, Washington, DC, December, pp. 46–60

Rosemann, M., De Bruin, T. and Power, B. (2006). A model to measure business process. In Jeston, J. and Nelis, J. (Eds.), *Business Process Management*. Butterworth-Heinemann, Oxford

Sarshar, M., Finnemore, M., Haigh, R. and Goulding, J. (1999). Spice: Is a capability maturity model applicable in the construction industry? Spice: A mature model. In Lacasse, M.A. and Vanier, D.J. (Eds.), *Durability of Building Materials and Components*. Institute for Research in Construction, Ottawa, ON, K1A 0R6, Canada, pp. 2836–2843

Savidas, A. (2009). *Your Guide to the NHS Infrastructure Maturity Model*. Informatics Directorate, Policy and Planning, Informatics Planning, Glasgow, United Kingdom

Sun, M., Vidalakis, C. and Oza, T. (2009). A change management maturity model for construction projects. In Dainty, A. (Ed.), *Proceedings 25 Annual ARCOM Conference*, Nottingham, England, 7–9 September, pp. 803–812

Van de Wetering, R. and Batenburg, R. (2009). A PACS maturity model: A systematic meta-analytic review on maturation and evolvability of PACS in the hospital enterprise. *International Journal of Medical Informatics*, 78: 127–140

Vivares, J.A., Sarache, W. and Hurtado, J.E. (2018). A maturity assessment model for manufacturing systems. *Journal of Manufacturing Technology Management*, 29(5): 746–767

Wendler, R. (2012). The maturity of maturity model research: A systematic mapping study. *Information and Software Technology*, 54(12): 1317–1339

6 Evaluation of relevant maturity models

Introduction

According to Newman (2017), few digital maturity models have been developed to help organisations take a holistic digital transformation approach. These models exist in the business world, manufacturing, and telecommunications. Since there is no digitalisation maturity model within the construction domain that could be adopted by construction organisations, understanding the different maturity models existing in other domains with rapt attention to the various capability areas (dimensions) is necessary. Through this, a clear view of the directions towards developing a digitalisation maturity model for construction organisations can be seen. This chapter, therefore, describes existing maturity models that are related to digitalisation. The first four models discussed in the chapter give insights into what the proposed construction digitalisation maturity model should entail regarding the dimensions and maturity levels. These four maturity models were selected because they all embraced the assessment of technology, which is a vital aspect of digitalisation, and adopted the five-maturity level concept of the original capability maturity model (CMM). The subsequent seven maturity models give the requirement of a digitalisation maturity model. Although they exist in other domains, they provide insights into what should be considered if an organisation seeks to attain digital transformation irrespective of the industry it belongs to.

Capability maturity model integration

The capability maturity model integration (CMMI) is an advancement of the CMM developed by the Software Engineering Institute (De Bruin *et al.*, 2005). According to Carnegie Mellon SEI (2005),

> [T]he CMMI Project was initiated to; build an initial set of integrated models, improve best practices from source models based on lessons learned, establish a framework to enable the integration of future models, and create an associated set of appraisal and training products.

The development of this model took the participation of over a 100 people in 30 organisations. The CMMI was developed to provide a structured view of process improvement across a software organisation. According to Paulk *et al.* (1993), a software organisation should be matured enough to manage the development of its software and the maintenance process required.

Furthermore, the process involved in the development and maintenance of this software must be clear to both existing and new staff of the organisation with work activities carried out strictly according to the planned process. This planned process must be workable and consistent with the production procedure within the organisation. Carnegie Mellon SEI (2005) further noted that the CMMI could help with the proper integration of organisations that are traditionally separate, set process improvement goals and priorities within an organisation, provide a road map for quality processes, and also provide a yardstick for appraising current practices.

CMMI can be represented in two ways, either staged or continuous. Although the data organisation and presentation are different in each representation, their content remains the same (Carnegie Mellon SEI, 2005). The stage representation considers five maturity levels related to the whole activity being assessed. For each level, some process areas that have to be improved to reach the specific maturity level are defined. Therefore, for an organisation at a particular level to get to the next level, it must enhance a subset of the predefined process areas, out of the whole, to attain such movement. The continuous representation, on the other hand, defines six capability levels, instead of maturity levels. A capability level represents a measure assigned to an isolated process area; thus, the maximum flexibility for organisations to choose which processes to point at is provided. Each process area has a different capability level, and the assessment of the whole of them makes up a capability profile. While the continuous representation uses predefined sets of process areas to define an improvement path for an organisation, the stage representation gives a sequence of improvements, each serving as a foundation for the next (Carnegie Mellon SEI, 2005; Macchi & Fumagalli, 2013). Based on this, Macchi and Fumagalli (2013) therefore conclude that the continuous representation offers the maximum flexibility for prioritising process improvements and aligning them with the business objectives. In the stage representation, a predefined path must be followed.

Carnegie Mellon SEI (2005) noted that four process areas (also referred to as dimensions and capability areas in this book) were adopted in the CMMI and they are:

Process management with variables such as organisational process focus, organisational process definition, organisational training, organisational process performance, and organisational innovation and deployment

Project management with variables such as project planning, project monitoring and control, supplier agreement management, integrated project management, risk management, integrated teaming, integrated supplier management, and quantitative project management

Engineering with variables such as requirements management, requirements development, technical solution, product integration, verification, and validation

Support with variables such as configuration management, process and product quality assurance, measurement and analysis, decision analysis and resolution, organisational environment for integration, and causal analysis and resolution

In terms of measuring maturity, the CMMI uses a five-level scale measurement system. These scales are described as 1- "Initial level" wherein the process is characterised as ad hoc, and occasionally even chaotic. The process is unpredictable, poorly controlled, and reactive; 2 – "Managed level" involves some planned project management processes. The process here is also often reactive. At this level, the necessary process discipline is in place to repeat earlier successes on projects with similar applications; 3 – "Defined level" wherein the process for both management and engineering activities is documented, standardised, and integrated into a standard process for the organisation. At this level, there is the use of an approved, tailored version of the organisation's standard software process for developing and maintaining software; 4 – "Quantitatively Managed level" which involves the detailed measurement of the software process and product quality. At this level, the software process and products are quantitatively understood and controlled; and 5 – "Optimising level" wherein continuous process improvement is enabled by quantitative feedback from the process and from piloting innovative ideas and technologies (Carnegie Mellon SEI, 2005; Paulk *et al.*, 1993).

Maintenance maturity assessment model

Macchi and Fumagalli (2013) developed a maintenance maturity assessment model geared towards measuring maintenance practices in an organisation. The model gave a scoring method for maturity assessment and a procedure for identifying the criticalities in maintenance processes with a view towards driving the improvement of the maintenance management system. In evaluating organisations' maintenance maturity, the model adopted three dimensions: managerial, organisational, and technological capabilities. It is believed that through these dimensions, the maturity level reached by a company can be analysed, in order to classify the criticalities in the organisation's maintenance processes. It is also assumed that a company can benchmark with the best companies of a reference sample using the developed assessment model. The management dimension assesses all process areas that deal with the planning and control cycle; this ranges from work order management to maintenance planning and budgeting. The organisational dimension involves all process areas with knowledge management and improvement of internal and external relationships. The technological dimension considers all processes relating to the support provided by diagnostic and prognostic tools, maintenance engineering tools, and other ICT tools. The aim of this dimension is not to only verify the presence of these tools but to also assess how they are effectively used in the organisation's practice.

In terms of maturity measurement in each of the three dimensions, the model, like most maturity models, draws from the CMMI concept. It adopts a five-level maturity scale. Level 1 is the initial level where the process is either poorly controlled or not controlled at all. Level 2 is the managed level wherein the process is partially planned, and performance analysis is mostly dependent on individual practitioners' experience and competences. Level 3 is the defined level wherein the process is planned, and semi-quantitative analyses are done periodically to define acceptable practices and management procedures. Level 4 is the quantitatively managed level where the process performance is measured, and the causes of special variations are detected. Level 5 is the last level known as the optimising level. At this level, the process is managed by ensuring continuous improvement.

Maturity assessment model for manufacturing systems

The maturity assessment model for manufacturing systems (MAMMS) was developed by Vivares *et al.* (2018) with the sole purpose of evaluating manufacturing system performance. The model was built following an action-research process, and the steps involved include the designing of the model, the maturity-level assessment in two manufacturing companies, and the validation of the model. There are two primary components; the first is 79 maturity assessed variables that were grouped into three categories (competitive priorities, manufacturing levers, and manufacturing strategic role). Competitive priorities include cost, quality, flexibility, innovativeness, deliveries, service, and environmental protection. These factors are geared towards giving the organisation a competitive advantage over other competitors. The manufacturing levers include human resources, structure and culture, sourcing and distribution, production planning and control, process technology, facilities, and management support subsystems. Lastly, the strategic role includes the role adopted in the organisation and continuous improvement.

The second component of the model is the five maturity levels used in evaluating the manufacturing system. These maturity levels are pre-infantile, infantile, industry average, adult, and world-class manufacturing. The model was tested and validated in two Columbian manufacturing companies. The model developed was considered a useful tool in establishing a manufacturing system's overall maturity level. The model deployment and validation results show that the MAMMS could be a tool valuable for decision-making in the management system. However, this model's significant limitation is its lack of details on its degree of relative importance as companies are often constrained (by market pressures, financial resources, etc.). Also, there is the need for improved understanding of a mechanism that can establish boundaries between maturity levels in specific quantitative variables. This can be in terms of deliveries, set-up time, inventory, accidents, unplanned downtime, anticipation, and the likes.

Telemedicine maturity model

Dyk and Schutte (2012) observed that the South African National Department of Health for many decades had recognised the inherent benefits of information and communication technology in the delivery of quality healthcare to rural areas. However, despite the generous funding and proven technology, only a few telemedicine systems have proved sustainable after the pilot phase. Based on this observation, the telemedicine maturity model (TMMM) that can help measure and manage the capability of a health system in the delivery of sustainable healthcare after the pilot phase of a telemedicine project was developed. This model was further validated within the context of the South African public health sector.

TMMM viewed maturity in five dimensions – machine (technology), money (finance), man (users), method (policy), and method (procedures). These maturity dimensions were derived from other existing maturity models (Broens *et al.*, 2007; Khoja *et al.*, 2007; Savidas, 2009; Van de Wetering & Batenburg, 2009). The technology aspect deals with telemedicine's needed technology, and it is deemed crucial in most telemedicine models. The finance category deals with the funding of telemedicine, which could either be a one-off investment from donors or pilot funds or through the existence of business models to ensure the continuation of the telemedicine endeavour. The policy aspect has to do with the existing policies that may or may not favour telemedicine, while the user's category is aligned with categories from other models, such as people and skills, learning readiness, and user acceptance. Lastly, the procedures are related to organisation and principles, standards, procedures, and guidelines (Dyk & Schutte, 2012). The TMMM adopts the conventional five maturity levels: initial level, managed level, defined level, measured level, and optimised level.

Strategic alignment maturity assessment model

Luftman (2000) developed a maturity model geared towards aligning businesses with IT. This was born out of the need for organisations to link technology with business due to the dynamic nature of business strategies and the ever-changing technologies that are constantly being developed (Papp, 1995; Luftman, 1996). According to Chan and Reich (2007), the word "alignment" is the stage whereby business and IT are in sync to achieve set corporate goals. Luftman (2000) stated that business-IT alignment could be described as the process of "*applying IT in an appropriate and timely way, in harmony with business strategies, goals and needs*". Based on the dynamic nature of the IT world, the need for the strategic alignment of business with IT has been emphasised by most researchers and professionals for many years (Boström & Celik, 2017). The strategic alignment maturity assessment model plays a crucial role in understanding the critical requirements for a maturity model that will not only help with the alignment of construction business with digital innovations but also assist in understanding where construction organisations stand currently within the digital world, and what they need to do to move beyond their current position and attain optimum digital transformation.

The strategic alignment maturity assessment model consists of six dimensions and five maturity levels, which are in line with the conventional five maturity levels of the CMM as noted earlier. This model's six dimensions include communication maturity, competence/value measurement maturity, governance maturity, partnership maturity, scope and architecture maturity, and skills maturity. The aspect of "communication" deals with the effective exchange of ideas and a clear understanding of a successful business strategy's necessities. It was observed that with the dynamic nature of the business world, ensuring continuous knowledge sharing across organisations is pertinent. Within the communication dimension, the variables are understanding of business by IT, understanding of IT by business, inter/intra-organizational learning, protocol rigidity, knowledge sharing, and liaison(s) effectiveness. The aspect of "competence/value measurement" is geared towards creating a balance "dashboard" that will adequately reveal the value of IT and how this will contribute to the business. It is believed that in most cases, the IT and business metrics in terms of value are not the same. Sometimes, IT organisations cannot paint the value their products will be adding to businesses. This creates a big decision issue for an organisation when it comes to adopting IT. The competence/value measurement consists of IT metrics, business metrics, balanced metrics, service level, agreements, benchmarking, formal assessments/reviews, and continuous improvement. "Governance" deals with the sharing of risk, conflict resolution, responsibilities as well as authority for resources among business and IT partners. This is achieved through formal discussion between both parties to review and allocate IT resources effectively. Within this dimension, variables include business strategic planning, IT strategic planning, reporting/organisation structure, budgetary control, IT investment management, steering committee(s), and prioritisation process.

It is believed that the relationship that exists between business and IT organisation plays a vital role in the strategic alignment of business with IT. This understanding gives birth to the "partnership" dimension, which includes variables such as the business perception of IT value, IT's role in strategic business planning, shared goals, risk, rewards/penalties, IT program management, relationship/trust style, and business sponsor. The scope and architecture dimension deals with the maturity of IT. This involves the extent to which IT enables a flexible infrastructure that is transparent to those involved in the organisation's activities. It also consists of the level at which IT evaluates and applies new technologies effectively and provides solutions that address specific customer's needs. The "scope and architecture" dimension has variables such as traditional, enabler/driver, standards articulation, architectural integration – functional organisation – enterprise – inter-enterprise, architectural transparency, flexibility, managing emerging technology. The last dimension is "skills" maturity with variables such as innovation, entrepreneurship, locus of power, management style, change readiness, career crossover, education, cross-training, social, political, trusting environment. This dimension deals mainly with the IT human resource considerations.

Using the CMM five-level maturity, the strategic alignment maturity assessment model is assessed on a five-level scale with 1 being the "initial/ad-hoc" level, 2 is the "committed level", 3 is the "established focus" level, 4 is the "improved or managed" level, and 5 is the "optimised" level.

Maturity model of digital strategising

By modifying Luftman's strategic alignment maturity assessment model, Boström and Celik (2017) came up with the maturity model of digital strategising (MMDS), which aimed to provide guidance for business owners in terms of framework and concepts that will aid proper strategising of their business. This model became necessary due to the need to move away from just aligning business with IT as in Luftman's (2000) model, to the point of assessing business in the current digital development being experienced. Also, it has been observed that currently, digital transformation is changing and reshaping business competition and at the same time impacting significantly on various aspects of businesses. In the digital world, IT is used to create value within organisations. However, these organisations are equally threatened by new competitors who might even be faster in using IT in their business delivery. Thus, strategic concerns are bound to arise within business managers regarding the digitalisation of their business and the effect this might have on their organisation, the industries, and the society at large (Boström & Celik, 2017; World Economic Forum, 2016). Bearing this in mind, the MMDS was developed to assist organisations in analysing and assessing their digital business strategy.

The model adopted Luftman's six dimensions but with a three-level maturity scale. These maturity levels are "IT strategising", "aligned strategising", and "digital strategising" (Boström & Celik, 2017). Also, some modifications were made to the dimensions to make it fit into the digital concept. These dimensions are communication maturity, value measurement maturity, leadership maturity, ecosystem maturity, technology maturity, and skill maturity. The aspect of "communication" involves the effective exchange of ideas and a clear understanding of a successful business strategy's necessities. Sledgianowski and Luftman (2005) have earlier noted that communication as a maturity dimension describes *"the mutual understanding between IT and business functions as well as the methods to support this understanding"*. This dimension consists of variables such as well-communicated and coherent digital strategy (Kane *et al.*, 2015), IT/business strategy is united into a digital strategy (Bharadwaj *et al.*, 2013), establish business and technical skills among IT and business people (McLaughlin, 2017), and boundary-spanning knowledge sharing (Lusch & Nambisan, 2015). The aspect of "value measurement" deals with the organisation's activities and their strategic IT choices, which reveals the value of IT to the business. The key variables relating to this dimension are the exploitation of data for decision-making generated through digital technologies (Bharadwaj *et al.*, 2013) and leverage digital options by investing in digital opportunities for the future (Coltman *et al.*, 2015). The "leadership" dimension was modified from Luftman's (2000) "governance" dimension since leadership better reflects the developing nature of digital business strategy. Sledgianowski and Luftman (2005) described this dimension as allocating strategic decisions on major IT projects. McKeown and Philip (2003) also stated that this dimension involves focusing on leadership issues, and this has become a critical success factor for digital business strategising. The variables under this dimension are embracing information-driven transparency through digital technologies (Bennis, 2013), building trust and commitment to the workforce for change (McKeown & Philip, 2003), and establishing risk as a cultural norm in the whole organisation (Kane *et al.*, 2015). The fourth dimension, which is "ecosystem" is also a modification of Luftman's (2000) "partnership" dimension. Variables within this dimension are establishing digital partnerships with external actors and manage relations (Bharadwaj *et al.*, 2013) and reacting fast to ecosystem changes and sensing environmental changes and responding to new IT initiatives (Bharadwaj *et al.*, 2013; Mithas *et al.*, 2013; Teece, 2007). The fifth dimension, "technology", was also a modification of Luftman's (2000) "scope and architecture". It refers to the effective usage of the large array of technologies emanating continuously and how the business process is driven by IT. The variables within this dimension include the flexibility of available technology towards the business, its environs (Bharadwaj *et al.*, 2013; McLaughlin, 2017), and the generation of competitive advantage through proper technology adoption (Bharadwaj *et al.*, 2013). The last dimension is "skills", which is seen as the management of IT human resource considerations. This includes upgrading digital skills through a proper alliance with external bodies (Bennis, 2013, Kane *et al.*, 2015) and full participation of employees in creating solutions and developing awareness of change (Kenny, 2006).

Forrester's Digital Maturity Model 4.0

Gill *et al.* (2016) in a bid to provide a road map for business organisations to achieve digital maturity developed a digital maturity model in 2014. Every year this model was tested for relevance and possible development through the inclusion of new ideas. The model was designed to accommodate three possible scenarios in business. These scenarios include: (a) overall digital transformation, which involves assessing the foundational aspects that matter to a company's overall digital transformation. This foundational aspect can be the support of company's executives for digital strategy, digital staffing, measuring of success within the organisation, and business functions/IT relationship effectiveness; (b) Digital marketing focus which involves the review of capabilities that are specific to a firm's digital marketing function; (c) Digital business focus which has to do with evaluating the support of digital transformation to sales and service interactions, including touchpoint integration and technology sophistication.

The model adopted four dimensions for determining digital maturity within an organisation: culture, technology, organisation, and insights. The "culture" dimension deals with a company's approach to digitally driven innovation and how it empowers employees with digital technology, while the "technology" dimension focuses on the company's use of emerging technologies in the delivery of its services. The "organisation" dimension involves the way and manner a company has aligned itself towards supporting digital strategy, governance, and execution. The last dimension, which is "insight", has to do with how well a company uses customer and business data to measure success and inform strategy.

Unlike most maturity models that measure maturity on a five-level scale, digital maturity model 4.0 measures maturity using four maturity levels. This further confirms Vivares *et al.*'s (2018) submission that although typical maturity models have been structured with five levels, some maturity models have adopted variant maturity levels other than the original five levels. The four maturity levels adopted in this model are Skeptics level – companies that are just beginning the digital journey; Adopters level – companies that are investing in digital skills and infrastructures; Collaborators level – companies that are breaking down traditional silos; and Differentiators – companies that are leveraging data to drive customer obsession. It is believed that after the critical evaluation of a company's current maturity level, further improvement must be carried out to help organisations move from the current maturity level to the next higher maturity level. Gill *et al.* (2016) suggested specific improvement measures that top managements can adopt in achieving this maturity feat.

A significant observation of this maturity model is that it gives a scale of range for measuring each maturity level. While the skeptic level falls within 0–33%, the adopter level falls within 34%–52%, the collaborator level falls within 53%–71%, and the differentiator level falls within 72%–84%. No maturity level was given for above 84%. However, it was stated that once the differentiator level has been attained, the next growth opportunity for these digital experts is to perfectly eliminate the separation between the digital and physical worlds. This implies that at 84, digital maturity has been attained, but these organisations can still achieve further maintenance and proper integration up to 100% (Gill *et al.*, 2016).

Digital maturity model for telecommunication providers

Valdez-de-Leon (2016) observed that digital transformation has taken over almost every industry and the telecommunications industry is no exception. It was noted that for Communication Service Providers (CSPs), this digital transformation started with the emergence of the so-called over-the-top services such as WhatsApp and Skype. However, despite such transformation, it was observed that there is paucity of frameworks and tools to help CSPs navigate such radical change. This knowledge led to the developing of a digital maturity model for telecommunication providers (DMMTP). It was noted that most CSP companies find it hard to attain digital transformation. In most cases, these companies rely on existing frameworks, including seminal works in IT-enabled business transformation and more recent developments in digital transformation practice for them to navigate their way through their desired digital transformation journey (Gerbert *et al.*, 2016).

Based on this knowledge, Valdez-de-Leon (2016) developed the digital maturity model using seven dimensions deemed crucial to CSPs. Like the Forrester's Digital Maturity Model 4.0, the DMMTP recognised the need to treat "technology" and "organisation" as key dimensions for digital maturity. While the technology dimension is seen as effective technology planning, deployment, integration, and use in supporting digital business, the organisation dimension looks at changes in culture, structure, training, and knowledge management that will enable the organisation to become a digital player. The remaining five dimensions recognised in this model are strategy, customer, ecosystem, operations, and innovation. Two of these dimensions (strategy and customer) are related to the insight dimension of the Forrester's Digital Maturity Model 4.0. Strategy dimension looks at the vision, governance, planning, and managing process that will support digital strategy execution, while the customer dimension looks at the new benefits created in customer experience through digital changes. All these are embedded in the "insight" dimension of the Forrester's Digital Maturity Model 4.0.

The significant difference between both models (Forrester's model and DMMTP), which can be attributed to the domains for which these models were developed, lies in introducing the ecosystem, operations, and innovation dimensions in the DMMTP. The ecosystem dimension was introduced to assess the partner ecosystem development and sustenance as a critical element for digital business, while the operations dimension was introduced to assess the capabilities of the service provision and increased maturity resulting from a more digitised automated and flexible operations. The innovation aspect covers the issue of using new approaches that are flexible and responsive to what is obtainable to have a more effective digital business.

In measuring maturity, DMMTP adopted a six-maturity level system, starting from zero to five. The zero level was set as a default level representing a state of inaction. Organisations at this level are yet to take steps towards digital transformation. Hence this level is known as the "Not started level". The first level is called the "initiating level", and organisations at this level have decided to move towards digital business and take steps towards digital transformation. The second level is known as the "enabling level" where organisations implement initiatives that form the foundation of their digital business. On completion of implementing initiatives at various dimensions of the business, an organisation can attain the third level, the "integrating level". At this level, the initiative of the organisation is integrated across the organisation to support end-to-end capabilities. Successful mastering of integrating an organisation's initiative will lead to the need to fine-tune these digital initiatives within the different dimensions to increase performance. This point of fine-tuning is the fourth maturity level called the "optimising level". The last stage of digital maturity is the "pioneering level", where the organisation tends to break new ground and advance the state of digital practice within the dimensions.

The DMMTP was developed as a tool meant to help gauge digital maturity at a particular point in time and help develop a vision and a road map for digital transformation. However, its major limitation is that the model is not prescriptive to suggest the best way to achieve the target state. Thus, the author suggested the development of complementary tools that will help define the best practices. This can be done through in-depth empirical evidence from more CSPs who are ready to embark on the journey of digital transformation and based on proper understanding and documentation of successes and failures of the digital process.

TM-forum digital maturity model

In developing a digital transformation blueprint for CSPs, the TM-Forum developed the digital maturity model (DMM) that can be used to attain digital transformation. Unlike the DMMTP that adopted seven dimensions in digital maturity, the DMM adopted five critical dimensions, 34 sub-components, and 175 variables. These dimensions include the "customer" with four sub-components, "strategy" with seven sub-components, "technology" that also has seven sub-components, "operations" with five sub-components, and the "culture/people/organisation" with four sub-components (Newman, 2017). These five dimensions are evident in the seven dimensions adopted in the DMMTP.

The customer dimension focuses on issues surrounding customer's satisfaction and experience with the organisation's products, behavioural patterns, and trust and perception. The strategy dimension deals with the complete strategy of the organisation in attracting and retaining customers. This includes all necessary management systems such as brand, environment, stakeholders, and strategic management. These two dimensions (customer and strategy) were combined into one in the Forrester's Digital Maturity Model 4.0, that is, the "insight" dimension, which tells how well a company uses customer and business data to measure success and inform strategy (Gill *et al.*, 2016). The technology dimension here deals with the organisation's ability to use technologies in the delivery of its products. It covers technological applications, data and analysis, networking, connected things, and technology architecture. The Forrester's Digital Maturity Model 4.0 and DMMTP also adopted technology as a crucial dimension. This is understandable as digitalisation comes with the organisation's ability to effectively utilise digital technologies in the delivery of their products and services. The operations dimension is more concerned with the process adopted in producing and delivering the organisation's product or service. The last dimension, which is the culture, people, and organisation, has to do with the organisation's culture towards digital innovation and their workforce enablement in the delivery of the organisation's product. This particular dimension was broken down into two separate dimensions in the Forrester's Digital Maturity Model 4.0. Culture and people were treated as one dimension called the "culture", and it dealt with the issue surrounding the company's approach towards adopting digital innovation, and the empowerment of its employees with digital technology. The "organisation" aspect was treated as a standalone dimension as it involves the way and manner a company has aligned itself towards supporting digital strategy, governance, and execution.

The model follows the conventional five maturity levels – initiating, emerging, performing, advancing, and leading. A typical description of these maturity levels are

> Initiating: discussions are early stage and beginning to be incorporated into some of our business operations; Emerging: discussions are advanced and

beginning to be incorporated into all daily operations; Performing: the organisation has set clear objectives and formulated a plan that is being followed throughout the company; Advancing: organisation is expanding on our plan and objectives to come up with new and innovative ideas to advance our capabilities in this area; Leading: organisation is considered a thought leader in this area, regularly leads industry discussions on the topic and has mastered this subject matter area.

International Data Corporation Digital Transformation Maturity Model

To provide quality IT guidance for businesses worldwide, the International Data Corporation (IDC), a Chinese-owned company that offers global, regional, and local expertise on technology and industry opportunities, developed a Digital Transformation Maturity Model (DTMM). Whalen (2015) observed that digital transformation is how organisations drive changes in their business models and ecosystems by leveraging digital competencies. Thus, the DTMM was designed to help businesses attain digital transformation that allows them to deliver better services to their customers. Over time, this model has been used as a benchmark to evaluate businesses' digital attainment around the world. In 2015, IDC on the feedback from 413 organisations in Europe described organisations in Europe to be at an Opportunistic (Digital Explorer) level in terms of digital maturity. These organisations ranked at the second stage of the five possible maturity levels (IDC, 2015).

The DTMM assessed digital transformation of business in five dimensions. It is believed that for a business to attain digital transformation successfully, it must be matured in the area of leadership, omni-experience, work-source, operating model, and information. The "leadership" dimension enables businesses to develop the vision for the digital transformation of products, services, and experiences optimised to deliver value to partners, customers, and employees. The "omni-experience" dimension describes an omnipresent and multidimensional ecosystem approach targeted towards intensifying experience excellence for products and services. The "work-source" dimension deals with the evolution of the way that businesses will achieve business objectives by effective sourcing, deployment, and integration of internal and external resources. The "operating model" dimension deals with making business operations more responsive and effective by leveraging digitally connected products, services, assets, people, and trading partners. The "information" dimension covers the focused approach of extracting and developing the value and utility of information relative to customers, markets, transactions, services, products, physical assets, and business experiences. Like the conventional maturity model, the DTMM adopts a five-level maturity scale. These five-maturity levels are; level 1 "Adhoc" (Digital Resister), level 2 "Opportunistic" (Digital Explorer), level 3 "Repeatable" (Digital Player), level 4 "Managed" (Digital Transformer), and level 5 "Optimised" (Digital Disrupter) (IDC, 2015).

Digital Library Maturity Model

The Digital Library Federation (1998) defined digital libraries (DL) as

> organisations that provide the resources, including the specialised staff, to select, structure, offer intellectual access to, interpret, distribute, preserve the integrity of, and ensure the persistence over time of collections of digital works so that they are readily and economically available for use by a defined community or set of communities.

Considering the overarching importance of DL in delivering and preserving knowledge within a community, the need for a maturity model is essential. Sheikhshoaei *et al.* (2018) noted that DL development is faced with diverse challenges which are complex. It was further stated that these complex problems could be tackled using a maturity model. Also, the development of a maturity model can increase performance and provide bases for achieving the desired DL. Based on this notion a Digital Library Maturity Model (DLMM) in Iran was designed through a qualitative approach, meta-synthesis, and a Delphi.

The model adopted the CMM five maturity levels design as a base model, and three main dimensions were identified. These dimensions are organisational governance, organisation–human, and technical content. The aspect of organisational governance deals with DL's strategic management and the continuous control of its function. This can be seen in the areas of planning of DL, DL architecture, ability to develop strategies to manage changes with the sole purpose of reducing challenges of DL, re-engineering of the processes involved in DL, creating targets, benchmarking other DLs, evaluating and monitoring the performance of DL, as well as measuring user satisfaction. Critical evaluation of these variables' intrinsic nature shows that they are similar to what is obtainable in the "strategy dimension" of the DMMTP developed by Valdez-de-Leon (2016) and the Forrester's Digital Maturity Model 4.0 developed by Gill *et al.* (2016). This is a pointer to the importance of having the right strategy in order to attain digital transformation.

The second dimension which has to do with the humans in the organisation deals with the workforce in the DL, the available infrastructure and support, and structures and processes available within the DL. These can be seen in the aspects of empowering employees, ability to attract and retain an experienced and qualified workforce, training for both employees and users, motivating employees, boosting team spirit among the organisation's workforce, promoting a DL culture, interacting with other DLs, proper need assessment, readiness to start a DL, as well as going the extra mile to offer services to DL users. The third aspect is the technical content of the DL. This dimension dwells on the content of the DL as well as issues surrounding IT infrastructure. Variables evident in this dimension are collecting and organising resources, improving the ability to search and accessibility of DL resources, managing information in a multilingual manner, improved usability, updating, archiving, and protection of resources, backing up DL using software and hardware systems as well as protecting them, ensuring information is secure, and improving user interface features. This dimension points out that the aspect of technology cannot be overlooked in the attainment of a digitally transformed organisation. Since the DLMM adopted the CMM as a base model, a five-level maturity was also adopted. These levels were described as level 1 "initial", level 2 "repeatable", level 3 "Defined", level 4 "managed", and level 5 "optimising".

Capability dimensions from related maturity models

A critical analysis of some of the aforementioned maturity models shows some distinct dimensions necessary for digital transformation, as shown in Table 6.1. While some of these dimensions tend to reoccur in most of these models, some other dimensions are peculiar to some specific models. This includes the ecosystem dimension which is evident only in the strategic alignment maturity assessment by Luftman (2000), MMDS by Boström and Celik (2017), and DMMTP by Valdez-de-Leon (2016), the policy and finance dimension is evident only in the TMMM by Dyk and Schutte (2012), while the customer dimension is evident only in the DMMTP by Valdez-de-Leon (2016) and DMM by TM Forum. However, while these dimensions stand alone in some of these models, there is evidence of their existence under other popular dimensions considered in other models. For example, the ecosystem dimension which has to do with partnering ecosystem development in the DMMTP and the finance dimension in the TMMM are evident in the strategy dimension in the TM Forum's DMM. Similarly, due to the similarities in some dimensions like process, procedures, and operations, they have been grouped together as shown in Table 6.1. While some models use a particular term to describe these dimensions, others use something different. However, their latent properties are the same, hence their grouping as one.

Building on the dynamic capability theory, discussed in Chapter 7, wherein process, position, and paths are vital in determining organisations dynamic capability and considering the particularity of construction organisations and the nature of their service delivery, this study considers the technology, process, people, and strategy dimensions for the development of the proposed maturity model. These selected dimensions are the most reoccurring dimensions, and a critical look at their sub-components shows that other dimensions such as culture, innovation, customers, ecosystem, policies, and finance all lie within these four major dimensions. Further to this, Nesensohn *et al.* (2014) has stated that maturity models typically describe an organisation's evolution over a defined period of time, of an organisation of people, technology, products, and processes.

A fifth dimension was introduced to assess construction organisations' maturity capability in partnering with other digital-oriented organisations within and outside the construction industry. Partnering has either been looked at under some other dimension or ignored in some cases. However, considering that the construction industry is lagging in its digital transformation, partnering with other organisations outside the industry regarding resources and expertise can help

Table 6.1 Existing maturity model dimensions

Authors	Technology	Organisation/ Culture/people	Strategy/ Monitoring	Process/ Operations	Innovation	Governance/ Leadership	Customer	Ecosystem	Policy	Finance
Luftman (2000)	✓	✓	✓			✓			✓	
Carnegie Mellon SEI (2005)	✓	✓	✓		✓					
Dyk and Schutte (2012)	✓	✓		✓					✓	✓
Macchi and Fumagalli (2013)	✓	✓	✓							
IDC (2015)		✓	✓	✓	✓					
Valdez-de-Leon (2016)		✓	✓	✓	✓	✓	✓	✓		
Gill et al. (2016)	✓	✓	✓							
Boström and Celik (2017)	✓	✓	✓			✓		✓		
De Carolis et al. (2017)	✓	✓	✓	✓						
Newman (2017)	✓	✓		✓			✓			
Sheikhshoaei et al. (2018)	✓	✓	✓			✓				
Vivares et al. (2018)	✓	✓	✓	✓	✓					
Total	11	12	10	7	3	4	2	4	1	1

Source: Author's compilation (2019)

construct organisations' digital transformation. In addition, partnering with other construction organisations with higher capability can lead to digital transformation for organisations with lesser digital capabilities.

Similarly, the project environment, which is considered as the area where construction activities are being carried out, is another important dimension that needs to be assessed individually. Considering the construction industry where the environment plays a significant impact in the successful delivery of construction (Akanni *et al.*, 2015; Aghimien *et al.*, 2020), individually assessing the digital maturity capability of construction organisations in relation to their environment is important. This assessment can be done based on the extent to which the external environment exerts pressure on the organisation to be digitally transformed.

Therefore, while four major dimensions (technology, people, process, and strategy) were drawn from Table 6.1 due to their reoccurrence in existing models, two additional dimensions (digital partnering and environment) that have not gained adequate attention in the reviewed models were added to create a more holistic model that can fit into the construction context. These dimensions, as well as their sub-attributes, are further discussed in Chapter 8.

Summary

This chapter assessed the relevant maturity models that can serve as a baseline in determining the necessary dimensions needed for a holistic digitalisation capability maturity model for construction organisations. The review of existing models revealed some significant capability areas wherein maturity need to be attained to achieve digital transformation. These identified areas are subsequently discussed in Chapter 8 of this book.

References

Aghimien, D.O., Aigbavboa, C.O., Oke, A.E. and Aghimien, L.M. (2020). Latent institutional environment factors influencing construction digitalization. *International Journal of Construction Education and Research*. doi: 10.1080/15578771.2020.1838973

Akanni, P.O., Oke, A.E. and Akpomiemie, O.A. (2015). Impact of environmental factors on building project performance in Delta State, Nigeria. *HBRC Journal*, 11: 91–97

Bennis, W. (2013). Leadership in a digital world: Embracing transparency and adaptive capacity. *Management Information Systems Quarterly*, 37(2): 635–636

Bharadwaj, A., El Sawy, O., Pavlou, P. and Venkatraman, N. (2013). Digital business strategy: Toward a next generation of insights. *Management Information Systems Quarterly*, 37(2): 471–482

Boström, E. and Celik, O.C. (2017). *Towards a Maturity Model for Digital Strategizing – A Qualitative Study of How an Organisation Can Analyse and Assess Their Digital Business Strategy*. IT Management Master Thesis submitted to the Department of informatics, UMEA Universitet, Sweden

Broens, T., Huis, I., Veld, R.M.H.A., Vollenbroek-Hutten, M.M.R., Hermens, H.J., Van Halteren, A.T. and Niewenhuis, L.J.M. (2007). Determinants of successful telemedicine implementations. *Journal of Telemedicine and Telecare*, 6(13): 303–309

Carnegie Mellon SEI. (2005). *Capability Maturity Model® Integration (CMMI) Overview.* Carnegie Mellon University, Pittsburgh, PA

Chan, Y.E. and Reich, B.H. (2007). IT alignment: What have we learned? *Journal of Information Technology*, 22(4): 297–315

Coltman, T., Tallon, P., Sharma, R. and Queiroz, M. (2015). Strategic IT alignment: Twenty-five years on. *Journal of Information Technology*, 30(2): 91–100

De Bruin, T., Freeze, R., Kaulkarni, U. and Rosemann, M. (2005). Understanding the main phases of developing a maturity assessment model. In Campbell, B., Underwood, J. and Bunker, D. (Eds.), *Australasian Conference on Information Systems (ACIS)*, Australia, New South Wales, Sydney, 30 November–2 December

De Carolis, A., Macchi, M., Negri, E. and Terzi, S. (2017). A maturity model for assessing the digital readiness of manufacturing companies. In Lödding, H. *et al.* (Eds.), *APMS 2017. IFIP Advances in Information and Communication Technology. (IFIPAICT, Vol 513).* Springer Verlag, Berlin, pp. 13–20

Digital Library Federation. (1998). A working definition of digital library. Available at: www.diglib.org/about/dldefinition.htm [accessed 19–11–2019]

Dyk, L.V. and Schutte, C.S.L. (2012). Development of a maturity model for telemedicine. *South African Journal of Industrial Engineering*, 23(2): 61–72

Gerbert, P., Castagnino, S., Rothballer, C., Renz, A. and Filitz, R. (2016). *Digital Engineering and Construction: The Transformative Power of Building Information Modelling.* The Boston Consulting Group, Boston, MA

Gill, M., VanBoskirk, S., Evans, P.F., Nail, J., Causey, A. and Glazer, L. (2016). The digital maturity model 4.0. Benchmarks: Digital business transformation playbook. Available at: www.forrester.com

International Data Corporation (IDC). (2015). Digital transformation: Benchmarking assessment IDC recommendations. Available at: www.idc.com

Kane, G., Palmer, D., Phillips, A. and Kiron, D. (2015). Is your business ready for a digital future? Available at: https://sloanreview.mit.edu/article/is-your-business-ready-for-a-digital-future/ [accessed 19–11–2020]

Kenny, J. (2006). Strategy and the learning organisation: A maturity model for the formation of strategy. *The Learning Organization*, 13(4): 353–368

Khoja, S., Scott, R., Casebeer, A., Mohsin, M., Ishaq, A. and Gilani, S. (2007). e-Health readiness assessment tool for healthcare institutions in developing countries. *Telemedicine Journal and e-Health*, 13(4): 425–431

Luftman, J. (1996). *Competing in the Information Age: Practical Applications of the Strategic Alignment Model.* Oxford University Press, New York

Luftman, J. (2000). Assessing business-IT alignment maturity. *Strategies for Information Technology Governance*, 4(14): 1–50

Lusch, R.F. and Nambisan, S. (2015). Service innovation: A service-dominant logic perspective. *MIS Quarterly*, 39(1): 155–175

Macchi, M. and Fumagalli, L. (2013). A maintenance maturity assessment method for the manufacturing industry. *Journal of Quality in Maintenance Engineering*, 19(3): 295–315

McKeown, I. and Philip, G. (2003). Business transformation, information technology and competitive strategies: Learning to fly. *International Journal of Information Management*, 23(1): 3–24

McLaughlin, S.A. (2017). Dynamic capabilities: Taking an emerging technology perspective. *International Journal of Manufacturing Technology and Management*, 31(1–3): 62–81

Mithas, S., Tafti, A. and Mitchell, W. (2013). How a firm's competitive environment and digital strategic posture influence digital business strategy. *MIS Quarterly*, 37(2): 511–536

Nesensohn, C., Bryde, D., Ochieng, E. and Fearon, D. (2014). Maturity and maturity models in lean construction. *Australasian Journal of Construction Economics and Building*, 14(1): 45–59

Newman, M. (2017). Digital maturity model (DMM): A blueprint for digital transformation (TM Forum White paper). Available at: www.tmforum.org

Papp, R. (1995). *Determinants of Strategically Aligned Organizations: A Multi-industry, Multi-perspective Analysis*. PhD Dissertation, Stevens Institute of Technology, Hoboken, NJ

Paulk, M., Curtis, B., Chrissis, M. and Weber, C. (1993). Capability maturity model for software, version 1.1. Available at: http://wwwsei.cmu.edu/pub/documents/93.reports/pdf/tR24.93.pdf

Savidas, A. (2009). *Your Guide to the NHS Infrastructure Maturity Model*. Informatics Directorate, Policy and Planning, Informatics Planning, Glasgow, United Kingdom

Sheikhshoaei, F., Naghshineh, N., Alidousti, S. and Nakhoda, M. (2018). Design of a digital library maturity model (DLMM). *The Electronic Library*, 36(4): 607–619

Sledgianowski, D. and Luftman, J. (2005). IT-business strategic alignment maturity: A case study. *Journal of Cases on Information Technology (JCIT)*, 7(2): 102–120

Teece, D. (2007). Explicating dynamic capabilities: The nature and micro foundations of sustainable enterprise performance. *Strategic Management Journal*, 28(8): 1319–1350

Valdez-de-Leon, O. (2016). A digital maturity model for telecommunications service providers. *Technology Innovation Management Review*, 6(8): 19–32

Van de Wetering, R. and Batenburg, R. (2009). A PACS maturity model: A systematic meta-analytic review on maturation and evolvability of PACS in the hospital enterprise. *International Journal of Medical Informatics*, 78: 127–140

Vivares, J.A., Sarache, W. and Hurtado, J.E. (2018). A maturity assessment model for manufacturing systems. *Journal of Manufacturing Technology Management*, 29(5): 746–767

Whalen, M. (2015). A digital transformation maturity model and your digital roadmap. A paper presented at the Agenda15 conference, Omni Amelia Island Plantation Resort, Amelia Island, Florida. Available at: http://www.agendaconference.com/wp-content/uploads/2014/12/Whalen_IDC.pdf [accessed 19–11–2020]

World Economic Forum. (2016). *Digital Transformation of Industries: In Collaboration with Accenture*. Digital Enterprise, WEF, Geneva Canton, Switzerland, pp. 1–45

7   Dynamic Capability Theory

Introduction

This book set out to develop a digitalisation capability maturity model that will help determine construction organisations' maturity in developing countries in terms of their capability to attain digital transformation. To determine construction organisations' capability in achieving this feat, this study builds on the Dynamic Capability Theory (DCT) viewpoint. It is believed that the construction industry in its nature is dynamic, and the construction market is not static (Navon, 2005). However, there lies some inherent capabilities within organisations that make them stand out from their competitors. Putting these capabilities into use can help attain the required competitive advantage. This chapter gives an overview of DCT, defines dynamic capabilities, and identifies the different dynamic capabilities needed by construction organisations for their digital transformation. It was noted that construction organisations must possess some first-order capabilities which are sensing, seizing, and transforming. However, these capabilities must be rooted in well-defined organisational and managerial processes, good positioning of the organisation's assets, and a clear evolutionary path. Therefore, these three areas (process, position, and paths) form the bases of the digitalisation capability maturity upon which other capability areas are rooted.

Dynamic Capability Theory

The Dynamic Capability Theory (DCT) is an extension of the Resource-based View (RBV) which has been described as a theoretical framework which influences the organisations' understanding of how competitive advantage can be achieved and sustained (Eisenhardt & Martin, 2000; Teece et al., 1997). According to Das and Teng (2000), RBV is based on the concept that organisations can only get a competitive advantage when they can effectively manage their internal resources. Past studies have noted that RBV assumes that within an organisation lies an abundance of heterogeneously distributed resources and differences of which persist over time (Amit & Schoemaker, 1993; Mahoney & Pandian, 1992; Wernerfelt, 1984). According to Eisenhardt and Martin (2000), this assumption of RBV have led to the theory of most researchers that once an organisation can have resources that have value, that is rare, that cannot be imitated nor substituted by another, such an organisation can achieve sustainable competitive advantage. This is because these resources can be used to implement new strategies that are value-oriented, and that cannot be easily duplicated by their competitors.

However laudable this view seems, it still has its shortcomings. Most significant is the submission of Teece *et al.* (1997) that the foundation of the RBV is not strong enough to support the attainment of a sustainable competitive advantage. This submission was based on the idea that though the RBV considers some mechanisms that can lead to the attainment of competitive advantage within an organisation, how these mechanisms go about sustaining this competitive advantage is omitted. Studies of Mosakowski and McKelvey (1997), Priem and Butler (2000), and Williamson (1999) have described this view as being "*conceptually vague and tautological*" due to its inability to adequately recognise the mechanisms that specifically contribute to the attainment of competitive advantage. Similarly, the view has been noted to have lacked empirical grounding (Priem & Butler, 2000; Williamson, 1999), nor is the attainment of a sustainable competitive advantage which it claims can be attained through the right resources possible in a competitive dynamic market (D'Aveni, 1994; Eisenhardt & Martin, 2000).

These shortcomings led to the development of the DCT by David Teece and Gary Pisano in 1994. According to Bleady *et al.* (2018), the development of DCT was deemed fit due to the inability of the RBV to "*interpret the development and redevelopment of resources and capabilities to address rapidly changing environments*". Teece *et al.* (1997) noted that the current business environment is not static as it has a continuous shifting topology. An organisation that will attain a sustainable competitive advantage over its competitors and survive in this rapidly changing environment must have a dynamic capability. Through this dynamic capability, its manager can effectively utilise both the internal and external competencies of the organisation that will tackle the changes within the business environment. Beske *et al.* (2014) stated that the introduction of DCT was to explicitly explain organisations' performance in dynamic business environments by dwelling on the capabilities of these organisations in obtaining a competitive advantage. Earlier studies have noted that maintaining a dynamic fit between what an organisation can offer and what the business environment requires has been a significant concern for most organisations regarding strategy and management (Learned *et al.*, 1965; Miles & Snow, 1978). To achieve this dynamic fit, an organisation must have dynamic capabilities that will increase the organisation's chances of survival and allow the organisation to grow (Helfat *et al.*, 2007).

Although some literature has characterised this theory as being vague and elusive (Kraaz & Zajac, 2001), its nature creates difficulty in determining the merits of its outcome (Winter, 2003; Zahra *et al.*, 2006), and being repetitive (Zollo & Winter, 2002), its ability to determine sustainable competitive advantage for organisations has been appreciated (Bleady *et al.*, 2018; Kuuluvainen, 2012).

Defining dynamic capabilities

Teece and Pisano (1994) have earlier clarified that the term "dynamic" is *"the capacity to renew competences to achieve congruence with the changing business environment; this is relevant in situations where time to market is critical, and the nature of competition is difficult to determine"*. "Capabilities" are referred to as *"the key role of strategic management in appropriately adapting, integrating and reconfiguring, internal and external organisational skills, resources, and functional competencies to match the requirements of a changing environment"*. Thus, by definition, Teece and Pisano (1994) described dynamic capabilities as *"the subset of the competencies/capabilities which allow the firm to create new products and processes, and respond to changing market circumstances"*. For more clarity, Teece *et al.* (1997) defined dynamic capabilities as *"the firm's ability to integrate, build and reconfigure internal and external competencies to address rapidly changing environments"*.

Following these definitions, several authors have proposed their definition of dynamic capabilities. The definitions proposed by the authors are in line with those of Teece's, and there is a general agreement in relation to what dynamic capabilities are. The definitions show that dynamic capabilities are organisational processes or routines by which organisations create or modify their firm's resources. Eisenhardt and Martin (2000) defined dynamic power as *"the firm's processes that use resources – specifically the processes to integrate, reconfigure, gain and release resources – to match and even create market change. It is the organisational and strategic routines by which firms achieve new resource configurations as markets emerge, collide, split, evolve, and die"*. Zollo and Winter (2002) used the keywords "learned", "stable pattern", and "systematically" in their definition to emphasise the point that dynamic capabilities are structured and continuous. They defined dynamic capability as *"learned and stable pattern of collective activity through which the organisation systematically generates and modifies its operating routines in pursuit of improved effectiveness"*. Wang and Ahmed (2007) emphasised that the continuous changes in an organisation's dynamic capability result from the constant changes that exist within the environment and the need for organisations to gain competitive advantage. They defined dynamic capability as *"a firm's behavioural orientation to constantly integrate, reconfigure, renew and recreate its resources and capabilities and, most importantly, upgrade and reconstruct its core capabilities in response to the changing environment to attain and sustain competitive advantage"*. While Helfat *et al.* (2007) described dynamic capability as the capacity of an organisation to purposefully create, extend, or modify its resource base, Ambrosini and Bowman (2009) defined it as variables which influence an organisation's resources and this, in turn, serves as the source of the firm's competitive advantage.

Torres *et al.* (2018) have further expanded upon the concept of dynamic capabilities and explain that they are structured and patterned processes of an organisation designed to reconfigure its ordinary capabilities towards successfully gaining competitive advantage by adapting to the changing environment. These organisational processes are unique to each organisation, as according to Schoemaker *et al.* (2018), it is difficult for other organisations to replicate and implement identical dynamic capabilities once thriving. The reasons provided are that dynamic capabilities present a challenge in terms of their development and deployment. Moreover, they are integrated with that particular organisation's processes, with its own culture, unique history, and problem-solving techniques (Schoemaker *et al.*, 2018). Therefore, while dynamic capabilities will help organisations withstand changes in the external environment, the success of dynamic capabilities depends on an organisation's internal environment.

Dynamic capabilities needed for construction digitalisation

Ambrosini and Bowman (2009) have stated that an organisation's dynamic capability influences its resources, which, in turn, serves as a source of the firm's competitive advantage. However, earlier studies (Learned *et al.*, 1965; Miles & Snow, 1978) note that maintaining a dynamic fit between what an organisation can offer and what the business environment requires (in terms of strategy and management) has been a significant concern for most organisations. Therefore, Teece (2007) advocated the need for some first-order capabilities (i.e., those needed in developing new second-order capabilities) such as sensing, seizing, and reconfiguring to maintain equilibrium between an organisation and its business environment. Past studies have also affirmed the quintessential importance of these first-order capabilities in organisations' survival in unpredictable environments and emerging markets (Eisenhardt & Martin, 2000; Schoemaker *et al.*, 2018; Zollo & Winter, 2002). Based on this understanding, these capabilities are discussed in relation to construction digitalisation.

Sensing capabilities

Sensing capabilities are systematic methods that deal with the acquiring of new information as well as sensing, shaping, categorising, and calibrating opportunities (Torres *et al.*, 2018). These systematic methods are regarded as the primary source of competitive advantage, which entail studying the external business (market) environment and factors that influence it, such as customers and competitors (Adeniran & Johnston, 2012; Alsemgeest *et al.*, 2017). These capabilities enable an organisation to have an advantage because the organisation can anticipate and understand customers' demands ahead of its competitors. Moreover, the sensing capability forms the basis of an organisation's ability to utilise market intelligence and the knowledge and awareness of changes in the market, which will enable better forecasting (Adeniran & Johnston, 2012). Furthermore, Endres (2018) alludes that the ability to sense threats and opportunities is an essential element for sustainable competitive advantage. An organisation's success depends upon identifying and developing opportunities and threats.

Teece (2007) opined that the dynamic nature of the business environments, technological opportunities, and competitor's activities are continually changing. Lots of opportunities present themselves to existing business entities and new ones. However, the profit stream of the existing business tends to be more at risk in this situation than those just coming into the business. Thus, they need to possess the ability to sense new opportunities and take advantage of them as quick as possible. Schoemaker *et al.* (2018) noted that organisations must sense market changes ahead of their rivals. Since Teece (2007) further described the sensing of opportunities as a "*scanning, creation, learning, and interpretive*" activity, for organisations to improve their ability to sense opportunities and detect threats within the business environment, Schoemaker *et al.* (2013) suggest that the monitoring of trends and uncertainties, and diverse systems of decision-making employed, must be strongly linked with tools for external scanning and scenario planning.

For construction organisations aiming to attain competitive advantage through the digitalisation of their services, the ability to sense game-changing opportunities and significant threats lurking within the environment is essential. Following Teece's (2007) suggestions, such construction organisations must put in place processes that will direct internal research and development (R&D) and select new digital technologies to help the organisation provide better services. Through R&D, the search for new digital technologies and processes can be done. This is important as it has been observed that construction organisations' inability to embrace new technological advancement and sponsor meaningful R&D has affected the service delivery of the construction industry in most developing countries (Aghimien *et al.*, 2018). Similarly, the organisations' managerial and organisational processes must be structured to tap into suppliers' and complementor's innovations. It has been observed that due to the dynamic nature of the business environment, uncertainty abounds (Quinton *et al.*, 2018). To scale through this uncertain environment, construction organisations can tap into suppliers' innovations or even mimic those organisations' innovations within the same industry (Gutierrez *et al.*, 2015). This process of mimicking others can lead to competitive advantage, improvement in technologies and innovations, and even development of innovative ideas based on what they have seen from others (Gutierrez *et al.*, 2015; Quinton *et al.*, 2018). Also, the organisation must put in place processes that will help identify target market segments, changing customers' needs, and customers' innovation. All this must be linked to a strong analytical system geared towards learning, sensing, filtering, shaping, and calibrating opportunities (Aghimien *et al.*, 2019; Froehlich *et al.*, 2017; Mendonça & Andrade, 2018; Teece, 2007; Yeow *et al.*, 2018). Based on the foregoing, the sensing capabilities needed can be summarised into the following:

- Putting processes in place to direct internal R&D
- Putting processes in place to tap into developments in exogenous science and technology
- Putting processes in place to identify target market needs, changing customer needs, and customer innovation
- Putting a process in place to help tap into suppliers' and complementor's innovations
- Having a robust analytical system

Seizing capabilities

Once organisations have acquired new information and knowledge on external and internal business factors and sensing and shaping opportunities, recent technological or market opportunity must be addressed through new processes, products, or services (Teece, 2007). Organisations must make strategic decisions on investing their resources and systematically design a business model to capitalise on the opportunities presented or to alleviate threats (Torres *et al.*, 2018). Seizing requires bringing productive resources into use in an innovative way for value capture from opportunities (Loux & Culié, 2016). Seizing can be seen as the process of organising necessary resources to meet the needs and opportunities identified by sensing actions to maximise the value from those actions (Aghimien *et al.*, 2019; Teece, 2014). This process is germane to aligning organisations to opportunities that have been identified using the organisation's sensing capabilities (Yeow *et al.*, 2018). Schoemaker *et al.* (2018) have noted that while it is essential to sense opportunity in an uncertain business environment (as in construction), timely implementation of newly identified systems and innovations is crucial. In the view of Teece (2007), the act of seizing transcends beyond just understanding of the new opportunities. It involves the actual decision-making of specific changes needed across the organisation to enjoy the benefits within these identified opportunities. Other studies have also described the seizing capabilities of an organisation as the organisational learning that is revealed through the ability to create knowledge within the organisation, obtain external knowledge, and assimilate this acquired knowledge for the creation of other necessary capabilities (Cepeda & Vera, 2007; Nyachanchu *et al.*, 2017). Thus, organisations must be conscious of the timely implementation of opportunities they have sensed within the construction environment. Also, the process of decision-making with regard to the implementation of these identified opportunities must be swift.

Based on the preceding, construction organisations seeking better competitive advantage through digital transformation must create a well-defined organisational structure, procedures, designs, and incentives geared towards seizing opportunities within the construction and digital environment. To achieve this, they must define their organisation's customer solution and business model based on the knowledge acquired from sensing. This can be done by selecting the right digital technology and service style, designing the avenue of revenue generation, selecting the target clients, and designing the right approach towards attaining value for both the organisation and its clients. Also, selecting the correct decision-making procedure and avoiding errors as much as possible is essential. Similarly, creating a clear boundary for the organisation's service is crucial to managing and controlling the organisations' service delivery. To achieve this, standardising unique assets of the organisation, controlling challenging assets, assessing appropriability, and recognising and managing co-specialised economies is essential. As earlier identified by Teece *et al.* (1997), these assets can be in the form of technologies, intellectual property, complementary assets, customer base, or even external relations of the organisation. Finally, in seizing opportunities, construction organisations must build loyalty and commitment. This they can achieve through quality leadership demonstration, effective communication within the organisation, with their clients, and with other project participants, and recognising non-economic factors, values, and culture (Aghimien *et al.*, 2019; Froehlich *et al.*, 2017; Haas, 2015; Teece, 2007). Based on this knowledge, the seizing capabilities needed can be summarised into the following:

- Well-defined customer solutions and business models
- Selection of the target clients
- Choosing the right digital technology and service style
- Planning and designing new structures and processes
- Designing the right approach towards attaining value for both the organisation and its clients
- Designing of the avenue of revenue generation
- Making routines for the selection of decision-making protocol
- Selecting the right decision-making procedure
- Routines to build loyalty and commitment
- Quality leadership demonstration
- Recognising non-economic factors, values, and culture
- Avoiding error

Reconfiguring

Schoemaker *et al.* (2018) observed that adapting to changes as they occur is not enough in an uncertain environment. Organisations may be required to reshape their activities and perhaps their ecosystems to enjoy new business models' benefits. Nyachanchu *et al.* (2017) noted that the act of reconfiguring refers to the organisation's ability to create and integrate capabilities from within and outside the organisation. Teece (2007) submitted that reconfiguration is the constant renewal of the organisation's asset, aligning these assets, co-alignment, and redeployment. This reconfiguration might require organisations to revamp their ways of conducting business, restructure units within the organisation, manage co-specialised assets, and create structures that allow knowledge development and good governance. Yeow *et al.* (2018) summarised this as the processes of restructuring the organisation's resources. Rindova *et al.* (2016) noted that through the act of reconfiguring, organisations could align existing resources with newly developed strategies, and at the same time adopt new resources to augment the shortfalls of the organisational resource base. Hence, reconfiguration capabilities can be seen as the capabilities to create and integrate (Nyachanchu *et al.*, 2017; Pavlou & El Sawy, 2011).

Based on the clamour for construction organisations to adopt innovative processes of delivering construction projects, reconfiguring construction organisation's processes through the digitalisation of existing processes and integrating new digital technologies to improve service delivery is necessary. Also, construction organisations that seek to attain competitive advantage through the digital transformation of their processes must be willing to embrace continuous alignment and realignment of their tangible and intangible assets with what is obtainable within the digital world. Such an organisation must decentralise its activities by embracing open innovation and developing its integration and coordination skills. Also, effective knowledge management is essential. Continuous learning of the use of evolving digital technologies, transferring of knowledge gained, and integrating such knowledge into the organisation's services is crucial (Froehlich *et al.*, 2017; Mendonca & Andrade, 2018; Teece, 2007). The reconfiguring capabilities needed can be summarised into the following:

- The digitalisation of existing processes by integrating new digital technologies
- Aligning organisational assets with digital requirements
- Decentralisation of organisational activities
- Continuous learning and knowledge management
- Co-specialisation
- Creation of new resources and processes

Hinging construction digitalisation in the dynamic capability theory

The DCT acknowledges three significant aspects that determine the attainment of competitive advantage for an organisation. These include the process (managerial and organisational), the position that depicts the types of assets within the organisation, and the paths that are the organisation's strategic directions (Teece & Pisano, 1994). Leonard-Barton (1992) earlier described dynamic capabilities within these three key areas. It was stated that "*the dynamic capabilities of an organisation reflect the organisation's ability to achieve new and innovative forms of competitive advantage given path dependencies and market positions*". On a simplified note, for an organisation to attain competitive advantage through dynamic capabilities of sensing, seizing, and reconfiguring, it must have a well-defined process, good positioning of its assets, and a clear path.

Process – in terms of organisational and managerial process involves coordination, integration, learning, and reconfiguration. Coordination is a static concept of learning which deals with transformation. It has been noted that managers are responsible for coordinating internal and external processes of their organisations, and this role is vital if a competitive advantage is to be attained. Learning is described as a dynamic process which involves repetition and experimentation that allows faster and more efficient execution of the task and in the process allows the unearthing of new opportunities. The last aspect of the process, that is, reconfiguration and transformation stems from the ability to sense and seize opportunities within the environment and transform organisational processes way ahead of competitors.

Therefore, construction organisations aiming to be digitally transformed and obtain a competitive advantage in the process must give adequate attention to coordination and integration of the right digital technology in their organisational process and services delivery. These organisations must also be willing to learn from these technologies' continuous use and be innovative in the processes. Investing in rigorous R&D in digital technology-related issues can help unearth new opportunities for these organisations.

Position – refers to the specific assets of the organisation that determines its strategic posture. These particular assets are unique to the organisation and are employed to gain a competitive edge over competitors. These specific assets can be in the form of unique knowledge and assets complementary to them and the organisation's reputation and relational assets. They can also be technological in terms of the acquiring of and using available technologies; financial assets; structural assets; market assets in terms of the position of the organisation's products in the market; customer base; and external relations of the organisation. The strategic position of an organisation in terms of its assets will influence the organisation's market share and profitability at any point in time.

Since digitalisation lies in the use of digital technologies, construction organisations must be willing to invest in the right technologies to aid their digital transformation. Positioning the organisation in an era of 4IR through adequate use of digital technologies is crucial not only to the digital transformation of the organisation but also to the organisation's profitability and market share.

Path – refers to the strategic alternatives that the organisation has and the presence or absence of increasing returns and attendant path dependencies (Aoki, 1990; Teece *et al.*, 1997). The destination of an organisation is an outcome of the organisation's positions and the path travelled. Therefore, the path of an organisation includes its path dependencies, organisation-specific technologies, and technological opportunities within the environment, and proper assessment of the process taken, the position determined by the organisation's specific assets, and the path.

From these three approaches and looking at the dynamic nature of the construction industry, it can be said that adopting the DCT view is a logical approach towards understanding the digital capabilities of construction organisations in developing countries. It is important to understand the process, position, and path within these construction organisations to attain digital transformation and get a better competitive advantage. Khin and Ho (2018) have noted that based on DCT, the digital capability could be considered an organisation's dynamic capability. It was further noted that the digital capability of an organisation complements the organisation's digital orientation since only organisations that have the skills to effectively manage new digital technologies will readily adopt them and make good use of these technologies in the delivery of new products. Benner (2009) has earlier proposed that the ability to respond to changes in terms of technology represents an element of dynamic capabilities. Therefore, based on the dynamic capabilities earlier identified (sensing, seizing, and reconfiguration), it can be deduced that organisations must be able to identify opportunities and threats that are available to them. Also, they must be able to seize these opportunities and use them to reconfigure their managerial and organisational process, their position in terms of assets, and their path.

Summary

This chapter explored the dynamic capability theory as a theory underpinning the proposed DCMM in the book. The chapter concluded that construction organisations must possess the capability to sense, seize, and transform their organisation by using their unique resources. However, these capabilities must be rooted in well-defined organisational and managerial processes, good positioning of organisation's assets, and a clear evolutionary path. Therefore, these three areas (process, position, and paths) form the digitalisation capability maturity bases upon which other capability areas are built. The chapter also linked the six dimensions earlier identified in Chapter 6 with these three major areas.

References

Adeniran, T.V. and Johnston, K.A. (2012). Investigating the dynamic capabilities and competitive advantage of South African SMEs. *African Journal of Business Management*, 6(11): 4088–4099

Aghimien, D.O., Aigbavboa, C.O. and Oke, A.E. (2018). Digitalisation for effective construction project delivery in South Africa. *Contemporary Construction Conference: Dynamic and Innovative Built Environment (CCC2018)*, Coventry University, Coventry, United Kingdom, 5–6 July, pp. 3–10

Aghimien, D.O., Aigbavboa, C.O. and Oke, A.E. (2019). Digitalisation of construction organisations in South Africa: A dynamic capability theory approach (2019). Advances in ICT in design, construction and management in architecture, engineering, construction and operations (AECO). *Proceedings of the 36th CIB W78 2019 Conference*, Northumbria University, Newcastle, United Kingdom, 18–20 September, pp. 74–83

Alsemgeest, L., Booysen, K., Bosch, A., Boshoff, S., Botha, S., Cunningham, P., Henrico, A., Musengi-Ajulu, S. and Visser, K. (2017). *Introduction to Business Management: Fresh Perspectives*, 2nd edition. Pearson, London, United Kingdom

Ambrosini, V. and Bowman, C. (2009). What are dynamic capabilities and are they a useful construct in strategic management? *International Journal of Management Reviews*, 11(1): 29–49

Amit, R. and Schoemaker, P.J.H. (1993). Strategic assets and organisational rent. *Strategic Management Journal*, 14(1): 33–46

Aoki, M. (1990). The participatory generation of information rents and the theory of the firm. In Aoki, M., Gustafsson, B. and Williamson, E. (Eds.), *The Firm as a Nexus of Treaties*. Sage, London, p. 2652

Benner, M.J. (2009). Dynamic or static capabilities? Process management and adaptation to technological change. *Journal of Product Innovation Management*, 26(5): 473–486

Beske, P., Land, A. and Seuring, S. (2014). Sustainable supply chain management practices and dynamic capabilities in the food industry: A critical analysis of the literature. *International Journal of Production Economics*, 152: 131–143

Bleady, A., Ali, A.H. and Ibrahim, S.B. (2018). Dynamic capabilities theory: Pinning down a shifting concept. *Academy of Accounting and Financial Studies Journal*, 22(2): 1–16

Cepeda, G. and Vera, D. (2007). Dynamic capabilities and operational capabilities: A knowledge management perspective. *Journal of Business Research*, 60(5): 426–437

D'Aveni, R.A. (1994). *Hyper-competition: Managing the Dynamics of Strategic Maneuvering*. Free Press, New York

Das, T.K and Teng, B. (2000). A resource-based theory of strategic alliances. *Journal of Management*, 26(1): 31–61

Eisenhardt, K.M. and Martin, J.A. (2000). Dynamic capabilities: What are they? *Strategic Management Journal*, 21(10–11): 1105–1121

Endres, H. (2018). *Adaptability Through Dynamic Capabilities: How Management Can Recognise Opportunities and Threats*, 1st edition. Springer Gabler, Switzerland

Froehlich, C., Bitencourt, C.C. and Bossle, M.B. (2017). The use of dynamic capabilities to boost innovation in a Brazilian chemical company. *Revista de Administração (São Paulo)*, 52(4): 479–491

Gutierrez, A., Boukrami, A. and Lumsden, R. (2015). Technological, organisational and environmental factors influencing managers' decision to adopt cloud computing in the UK. *Journal of Enterprise Information Management*, 28(6): 788–807

Haas, A. (2015). Micro-foundations of dynamic capabilities. The diverse roles of boundary spanners in sensing/shaping and seizing opportunities. *Proceeding of XXIVème conférence annuelle de l'Association Internationale de Management Stratégique* – AIMS, Paris, France, June

Helfat, C.E., Finkelstein, S., Mitchell, W., Peteraf, M., Singh, H., Teece, D.J. and Winter, S.G. (2007). *Dynamic Capabilities- Understanding Strategic Change in Organizations*. Blackwell Publishing, Oxford

Khin, S. and Ho, T.C.F. (2018). Digital technology, digital capability and organisational performance: A mediating role of digital innovation. *International Journal of Innovation Science*, 11(2): 177–195

Kraaz, M. and Zajac, E. (2001). How organisational resources affect strategic change and performance in turbulent environments: Theory and evidence. *Organization Science*, 12(5): 632–657

Kuuluvainen, A. (2012). How to concretise dynamic capabilities? Theory and examples. *Journal of Strategy and Management*, 5(4): 381–392

Learned, E., Christensen, R., Andrews, K. and Guth, W. (1965). *Business Policy – Text and Cases*. Irwin, Homewood, IL

Leonard-Barton, D. (1992). Core capabilities and core rigidities: A paradox in managing new product development. *Strategic Management Journal*, 13: 111–125

Loux, P. and Culié, J.D. (2016). Revisiting dynamic capabilities and their micro foundations: A study in the context of business schools. *32nd annual IMP Conference*, Poznan, Poland

Mahoney, J.T. and Pandian, J.R. (1992). The resource-based view within the conversation of strategic management. *Strategic Management Journal*, 13(5): 363–380

Mendonça, C.M.C.D. and Andrade, A.M.V.D. (2018). Dynamic capabilities and their relations with elements of digital transformation in Portugal. *Journal of Information Systems Engineering & Management*, 3(3): 1–8

Miles, R.E. and Snow, C.C. (1978). *Organizational Strategy, Structure, and Process*. McGraw-Hill, New York, NY

Mosakowski, E. and McKelvey, B. (1997). Predicting rent generation in competence-based competition. In Heene, A. and Sanchez, R. (Eds.), *Competence-Based Strategic Management*. Wiley, Chichester, pp. 65–85

Navon, R. (2005). Automated project performance control of construction projects. *Automation in Construction*, 14: 467–476

Nyachanchu, T.O., Chepkwony, J. and Bonuke, R. (2017). Role of dynamic capabilities in the performance of manufacturing firms in Nairobi county, Kenya. *European Scientific Journal November*, 13(31): 438–454

Pavlou, P.A. and El Sawy, O.A. (2011). Understanding the elusive black box of dynamic capabilities. *Decision Sciences*, 42(1): 239–273

Priem, R.L, and Butler, J.E. (2000). Is the resource-based 'view' a useful perspective for strategic management research? *Academy of Management Review*, 26(1): 22–40

Quinton, S., Canhoto, A., Molinillo, S., Pera, R. and Budhathoki, T. (2018). Conceptualising a digital orientation: Antecedents of supporting SME performance in the digital economy. *Journal of Strategic Marketing*, 26(50): 427–439

Rindova, V.R., Martins, L. and Yeow, A. (2016). The hare and the fast tortoise: Dynamic resource reconfiguration and the pursuit of new growth opportunities by yahoo and google (1995–2007). In *Resource Redeployment and Corporate Strategy (Advances in Strategic Management, Vol. 35)*, Emerald Group Publishing Limited, Bingley, pp. 253–284

Schoemaker, P.J.H., Day, G.S. and Snyder, S.A. (2013). Integrating organisational networks, weak signals, strategic radars and scenario planning. *Technological Forecasting & Social Change*, 80(4): 815–824

Schoemaker, P.J.H., Heaton, S. and Teece, D.J. (2018). Innovation, dynamic capabilities, and leadership. *California Management Review*, 6(1): 15–42

Teece, D. (2007). Explicating dynamic capabilities: The nature and microfoundations of sustainable enterprise performance. *Strategic Management Journal*, 28(8): 1319–1350

Teece, D.J. (2014). The foundations of enterprise performance: Dynamic and ordinary capabilities in an (economic) theory of firms. *Academy of Management Perspectives*, 24(4): 328–352

Teece, D.J. and Pisano, G. (1994). The dynamic capabilities of firms: An introduction. *Industrial and Corporate Change*, 3(3): 537–556

Teece, D.J., Pisano, G. and Shuen, A. (1997). Dynamic capabilities and strategic management. *Strategic Management Journal*, 18(7): 509–533

Torres, R., Sidorova, A. and Jones, M.C. (2018). Enabling firm performance through business intelligence and analytics: A dynamic capabilities perspective. *Information & Management*, 55(7): 822–839

Wang, C.L. and Ahmed, P.K. (2007). Dynamic capabilities: A review and research agenda. *International Journal of Management Reviews*, 9(1): 31–51

Wernerfelt, B. (1984). A resource-based view of the firm. *Strategic Management Journal*, 5(2): 171–180

Williamson, O.E. (1999). Strategy research: Governance and competence perspectives. *Strategic Management Journal*, 20(12): 1087–1108

Winter, S. (2003). Understanding dynamic capabilities. *Strategic Management Journal*, 24(10): 991–995

Yeow, A., Soh, C. and Hansen, R. (2018). Aligning with new digital strategy: A dynamic capabilities approach. *The Journal of Strategic Information Systems*, 27(1): 43–58

Zahra, S., Sapienza, H.J. and Davidsson, P. (2006). Entrepreneurship and dynamic capabilities: A review, model and research agenda. *Journal of Management Studies*, 43(4): 917–955

Zollo, M. and Winter, S.G. (2002). Deliberate learning and the evolution of dynamic capabilities. *Organization Science*, 13(3): 339–351

Part IV

Conceptual perspective of construction digitalisation capability maturity

Part IV of this book is focused on the conceptual perspective of construction digitalisation capability maturity. This part consists of two chapters. Chapter 8 focuses on the model's dimensions and sub-attributes, while Chapter 9 explores the conceptualised model in a developing country using expert opinion.

8 Conceptualisation of the construction digitalisation capability maturity

Introduction

This chapter gives the conceptual perspective of the digitalisation capability maturity model (DCMM). The chapter provides a description of the different dimensions in the DCMM and the sub-attributes measuring each dimension. The first four dimensions (technology, people, process, and strategy) were extracted from maturity models reviewed earlier, while the remaining two dimensions (digital partnering and environment) were gaps identified in existing models. These gaps were deemed necessary due to the slow adoption of digital technologies in the construction industry and the fact that the environment affects how the product of the industry is delivered. Based on these descriptions, the chapter presented the conceptualised DCMM and gave the structural model's specification and justification. Furthermore, the maturity level of the DCMM and the expected outcomes of the model are clearly defined in the chapter. At the end the conceptualised DCMM is theorised, and a brief description of how the model can be applied in a construction organisation is given.

Variable selection for digitalisation capabilities

This book set out to develop a digitalisation capability maturity model (DCMM) for construction organisations. As stated in Chapter 7, the development of this model relied on the Dynamic Capability Theory (DCT) concept since the digital world in its nature is dynamic and the construction market is not static (Aghimien et al., 2019; Navon, 2005). It is believed that there are some capabilities which an organisation is supposed to have in the context of the process, the path, and the position of the organisation that can help them to correctly sense, seize, and transform their activities and in the process attain competitive advantage. Similarly, for a construction organisation to be digitally transformed, the process of delivery of organisation's services as well as managerial processes, the position of the organisation in terms of digital technology, customer base, and other necessary resources, and the path in terms of the strategic alternatives of the organisation are important. Following the review of the existing literature in Chapter 6 on related maturity models which exist in the general business world, manufacturing, and telecommunications, as a result of the non-availability of a digitalisation maturity model within the construction industry that could be adopted for this study, certain dimensions (capability areas) and sub-attributes were revealed. These dimensions and their sub-attributes are discussed in this chapter.

Dimensions from existing models

This section discusses the four major dimensions that reoccurred in existing models reviewed earlier. These dimensions are considered significant to the digitalisation of construction organisations as they have been explored significantly in other industries as adjudged key capability areas for organisations seeking digital transformation. These dimensions are technology, people, process, and strategy.

Technology

Teece *et al.* (1997) described the availability of needed technology and infrastructure as a "position" dimension for any organisation seeking dynamic capability. The technology aspect of digitalisation is a reoccurring dimension in the models reviewed. This is because technology plays a vital role in the digital transformation of any organisation or industry. The need to adopt technologies in the delivery of construction services has been stressed in previous research (Agarwal *et al.*, 2016; Castagnino *et al.*, 2016; Oke *et al.*, 2018). Similarly, research on the benefits of technologies on the construction industry's overall productivity abounds (Astebro, 2002; Goodrum & Haas, 2004; Sepasgozar & Bernold, 2013). According to Eidhoff *et al.* (2016), the critical factors determining whether an organisation will go the digital route or not are embedded in the technological and environmental dimensions. Gatignon and Xuereb (1997) also submitted that concrete technological orientation is needed for a company to attain superior innovation over its counterparts. This technological orientation can be seen in the organisation's readiness to adopt new technologies and its attitude towards technological changes within its environment. According to Quinton *et al.* (2018), digital technology's adoption creates an avenue to achieve value for the organisation. It has earlier been observed that most organisations adopt these digital technologies to improve communication with customers and get the best out of proper processing of their information, increase the efficiency of the organisation's operations, as well as achieve organisational growth in the process (Borges *et al.*, 2009; Bhaskaran, 2013; Harrigan *et al.*, 2011; Quinton *et al.*, 2018).

Khin and Ho (2018) concluded that since digital technologies abound and their production is being improved upon by the day, companies should strive towards adopting these technologies so that innovative digital solutions can be obtained. According to Telkom's report in 2016, for digitalisation to be attained within an organisation, identifying what and how new technology will impact such an organisation's business activity is imperative. By doing this, the existing business process can be digitised using electronic channels, content, and transactions. However, it was pointed out that it is important to create a balance between electronic capabilities and traditional business when digitising certain business processes (Telkom, 2016). This implies that even though these technologies may exist for use, it requires the management's initiative to be put into effective usage. This is important because past research has noted that while technology might play a prominent role in the digital transformation of an organisation, the technology dimension of the digital transformation process is only important in terms of how well an organisation can restructure the way it practices technology adoption while creating new products and services to drive competitive advantage (Kane et al., 2015; McLaughlin, 2017; Yoo et al., 2010).

From the digital business viewpoint, Bharadwaj et al. (2013) observed that organisational strategy formulated and executed by leveraging digital resources (such as digital technologies) to create differential value births the concept of digital business strategy. Yoo et al. (2012) opined that a critical factor influencing the ongoing inclusion of digital strategy in business is digital technology and its unique properties, enabling companies to innovate in new and different ways. Bharadwaj et al. (2013) further acknowledged digital technologies as digital resources that can aid competitive advantage within any organisation. It was stated that "*firms become more digital and rely on information, communication, and connectivity functionality*" when adopting digital technologies.

Newman (2017) describes the technology dimension as the successful attainment of digital strategy through the creation, processing, storing, securing, and exchanging of data to meet clients' needs. All these are made possible by adopting variables relating to technology architecture (which includes the availability of all technological infrastructure needed), security, application, data and analysis, connected things, network, and delivery governance. According to Valdez-de-Leon (2016), this technological aspect of digital transformation is effective technology planning, deployment, integration, and use. Luftman (2000) has earlier described this dimension as the effective adoption of emerging technologies and how information technology is driving the business processes. Macchi and Fumagalli (2013) noted that technology capability of an organisation (with specific reference to maintenance) is crucial as the organisation needs to be able to adopt technology for monitoring, diagnostics and prognostics system; computerised maintenance management system; and reliability and maintenance engineering system. Similarly, in construction, there is the need for technology planning, development, integration, and use.

Zhou and Wu (2010) noted that technology capability deals with getting the right digital technologies, recognising new digital opportunities, taking steps towards attaining digital transformation by mastering modern digital technologies, and the development of organisation's products using the right technology. Boström and Celik (2017) viewed this dimension as the effective usage of different digital technologies and the extent to which the business process is driven by IT. This includes the flexibility of available technology towards the business and its environs and the generation of competitive advantage through proper technology adoption (Bharadwaj et al., 2013; Boström & Celik, 2017; McLaughlin, 2017). Sheikhshoaei et al. (2018) submitted that this dimension involves the efficient use of digital technology to gather resources, protect resources and put resources into effective use. According to Gill *et al.* (2016), this technology dimension can be seen as

> having fluid technology budget that allows adjustment in priorities, a collaboration between marketing and technological resources to create a road map for digital technology usage, having flexible, iterative and collaborative approach toward the development of technology, leveraging modern technology architectures to increase speed and flexibility of product delivery, measuring technology team through the outcomes of the business and not just system up-time, using customer experience assets to direct technology design, and promoting employee innovation through the use of digital technologies.

Andriole (2016) describes the 15 compulsory technology capabilities that an organisation that wants to survive in the current digital world must have. The study noted five crucial technologies, five management best practices and five essential soft skills. The technologies include cloud computing, strong analytics, digital security, digital media, and embracing and optimising emerging technologies. Similarly, Bibby and Dehe (2018) from the perspective of 4IR maturity termed the technology dimension as "factory of the future" with digital technologies such as 3D printing, cloud computing, IoT and CPS, big data, sensors, autonomous robots, and e-value chain listed as some of the critical technologies that organisations need to gain maturity in their usage.

From the aforementioned, it is clear that technology capability of a construction organisation aiming for digital transformation surpasses just the acquiring of the required digital technology and entails every process surrounding its effective usage in the organisation's service delivery. Based on the preceding, the 15 sub-attributes to measure a construction organisation's technology capability are listed in Table 8.1.

Table 8.1 Technology dimension of DCMM

Technology	Authors
Technology planning	Valdez-de-Leon (2016)
Emerging technologies optimisation	Andriole (2016)
Big data analytics	Andriole (2016), Newman (2017)
Additive manufacturing	Bibby and Dehe (2018)
Cloud computing/Connected things (IoT, IoE, IoS)	Andriole (2016), Newman (2017)
Standard IT systems	Newman (2017), Vivares *et al.* (2018)
Design modelling and simulations (BIM usage)	Gill *et al.* (2016), Newman (2017), Zhou and Wu (2010)
Design simulation and virtualisation (virtual, augmented and mixed reality usage)	Khin and Ho (2018)
Robotics and automation in construction	Gill *et al.* (2016), Newman (2017), Zhou and Wu (2010)

Technology	Authors
Digital media marketing	Andriole (2016), Gill *et al.* (2016)
Cybersecurity in organisations	Andriole (2016), Sheikhshoaei *et al.* (2018), Newman (2017)
Technology governance (Having principles governing technology delivery in place)	Sheikhshoaei *et al.* (2018), Newman (2017)
Continuous technology upgrades	Dyk and Schutte (2012)
Flexible technology usage	Bharadwaj *et al.* (2013), Boström and Celik (2017), Gill *et al.* (2016), McLaughlin (2017)
The synergy of technology with other organisation's resources	Gill *et al.* (2016)

People

Every activity within the construction industry is championed by people, guided by some specific culture within their organisation. Thus, people's role in the digital transformation of construction organisations cannot be overlooked if significant digital transformation is to be attained. The "people" dimension is evident in the DCT under the "position" dimension. The organisation's position depicts the types of assets within the organisation (people inclusive) (Teece & Pisano, 1994). From past related models reviewed in Chapter 6, people, culture, and organisations have been measured and grouped using similar variables. The selection of the name "people" for this dimension is based on the need for clarity of this dimension's function and what it should assess.

From the perspective of organisation, Macchi and Fumagalli (2013) submitted that the relationships between functions within the organisation, relationships with third parties (outsourcing), and empowerment of personnel are crucial sub-attributes to consider. Looking at this from the digitalisation viewpoint, it can be said that in attaining digital transformation in an organisation, there is the need for a digital relationship between functions in the organisation, digital relationships with third parties, and digital empowerment of personnel. Although Gill *et al.* (2016) in their digital maturity model separated the "organization" dimension from the "culture" dimension, considerable similarity can be seen in both dimensions. While the organisation dimension involves how a company has aligned itself towards supporting digital strategy, governance, and execution, the culture dimension deals with a company's approach to digitally driven innovation and how it empowers employees with digital technology.

The TM forum DMM considers "people", "organization", and "culture" as a single dimension that includes the digital culture within an organisation, organisational design and talent management, leadership and governance, and workforce enablement (Newman, 2017). Quinton *et al.* (2018) noted that senior management within an organisation plays a significant role in adopting technologies. This is because their level of knowledge in IT-related issues largely influences their behaviour, either supporting digital transformation or opposes it. Similarly, it affects their understanding of the inherent benefits of adopting digital technologies to the organisation. Research has also shown that having a positive attitude towards change in general, ability to take risk, proactive nature, readiness to challenge the status quo, and readiness to innovate, all have a significant effect on the adoption of digital technologies and the subsequent attainment of digital transformation (Grant *et al.*, 2014; Jones *et al.*, 2013; Peltier *et al.*, 2009, 2012).

Luftman's (2000) strategic alignment model viewed issues relating to people under its "skill" dimension. This dimension encompasses innovation within the organisation, entrepreneurship, locus of power, leadership/management style, change readiness, career crossover, education, cross-training, social, political, trusting environment. Building on Luftman's perception, Boström and Celik (2017) viewed the skill dimension as the degree of IT human resources management. This can be seen from the viewpoint of upgrading digital skills through a proper alliance with external bodies as observed by Bennis (2013) and Kane *et al.* (2015) and full participation of employees in terms of creating solutions and developing awareness of change (Kenny, 2006). According to Sheikhshoaei *et al.* (2018), the people dimension of digital transformation encompasses the empowering of employees, organisations' ability to attract and retain an experienced and qualified workforce, training for both employees and users, motivating employees, boosting team spirit among the organisation's workforce, promoting a digital culture, proper need assessment, etc. Bibby and Dehe (2018) from the perspective of 4IR maturity highlighted the importance of people and culture. This dimension requires construction organisations to have innovative openness and continuous improvement culture to propel technology usage in the organisation. Based on the preceding, the eleven sub-attributes to measure people capability of construction organisation in relation to digitalisation is listed in Table 8.2.

Table 8.2 People dimension of DCMM

People	Authors
Digital technical know-how of personnel	Gill *et al.* (2016), Luftman (2000), Sheikhshoaei *et al.* (2018)
Digital knowledge management by the organisation	Jones *et al.* (2013), Peltier *et al.* (2009), Quinton *et al.* (2018)
Reskilling of workforce	Gill *et al.* (2016), Newman (2017), Sheikhshoaei *et al.* (2018)
Continuous learning of personnel	Day (2011), Gill *et al.* (2016), Luftman (2000), Quinton *et al.* (2018), Sheikhshoaei *et al.* (2018)
Top management support	Gill *et al.* (2016), Quinton *et al.* (2018)
Digital culture within an organisation	Sheikhshoaei *et al.* (2018)
Organisation's positive change attitude	Luftman (2000), Peltier *et al.* (2012), Quinton *et al.* (2018)
Digital empowerment of personnel	Gill *et al.* (2016), Newman (2017), Sheikhshoaei *et al.* (2018)
Personnel's innovativeness	Luftman (2000), Jones *et al.* (2013)
Attracting digital talent	Sheikhshoaei *et al.* (2018)
Retaining digital talent	Sheikhshoaei *et al.* (2018)

Process

The process dimension is also referred to as "operations" in some reviewed models. Teece *et al.* (1997) referred to this aspect of the dynamic capability of any organisation as the organisational and managerial processes that deal with the coordination of activities within the organisation to deliver better product/services. Valdez-de-Leon (2016) introduced this dimension to assess service provision capability and increased maturity resulting from more digitised automated and flexible operations. Newman (2017) noted that organisations need to have good digital management at the operation stage in terms of automation, integration, and change. In construction, this can be likened to the need for good digital management at the planning/design, construction and operating phases of construction project delivery. Agile change management and integrated service management are required at every stage and in every organisation's function.

Similarly, there should be real-time insights and analytics and smart process management of the organisation's operations (Newman, 2017). De Carolis *et al.* (2017) observed that the operational aspect is crucial in the digital maturity of manufacturing companies. This tells the level of digital technology put into delivery of products within the industry. International Data Corporation (IDC) (2015) noted that organisations need to be able to make their business operations more responsive and effective by leveraging digitally connected products, services, assets, people, and trading partners. Boström and Celik (2017) mention "value measurement" as a crucial aspect of digital maturity. This is seen as the value adoption of IT to the business at every strategic point. Gebhardt *et al.* (2006) have noted that adding value to the business is complex, and it cuts across different functions within the organisation. Thus, in creating value for the organisation's product, applying technologies to every aspect of the business is essential. This dimension is measured by exploiting data for decision-making generated through digital technologies (Bharadwaj *et al.*, 2013; Boström & Celik, 2017) and leveraging digital options by investing in digital opportunities for the future (Boström & Celik, 2017; Coltman, 2015).

In construction, the construction process from design to the main construction and operation also needs to be assessed for its digital ability. At the design stage, digital technologies can be applied in measurement and acquiring of data (Vähä *et al.*, 2013), merging design by different construction participants, identifying interdependencies and clashes in design, achieving virtual migration of physical structures, and more data-driven design (Castagnino *et al.*, 2016; Gerbert *et al.*, 2016), effective cost estimation (Ehlhardt, 2014) e-tendering and e-procurement (Berger, 2016; Ibem & Laryea, 2016; Solanke & Fapohunda, 2015). During construction, digital technologies can be applied in real-time data sharing, integration and coordination, data-driven construction planning and lean execution, automated and autonomous construction, new fabrication methods, and rigorous construction monitoring and surveillance (Gerbert *et al.*, 2016) Furthermore, digital technologies can aid project monitoring, checking for deterioration, facilitating predictive maintenance, and continually updating a central database (Castagnino *et al.*, 2016), reduction of the total life cycle cost of construction projects (Delgado *et al.*, 2017), and enhance maintenance of structures (Gerbert *et al.*, 2016).

Table 8.3 Process dimension of DCMM

Process	Authors
Digitalised planning process	PMBOK (2017)
Automated construction process	PMBOK (2017), Valdez-de-Leon (2016)
Digitalised schedule management	PMBOK (2017)
Digitalised resource management	Newman (2017), PMBOK (2017)
Management of integrated service	Newman (2017), PMBOK (2017)
Value delivery and quality management	Boström and Celik (2017), PMBOK (2017)
Digitalised stakeholder management	PMBOK (2017)
Digitalised scope management	PMBOK (2017)
Digitalised cost management process	PMBOK (2017)
Digitalised procurement process	PMBOK (2017)
Digital risk management	PMBOK (2017)
Change management	Newman (2017)
Communication management	PMBOK (2017)

The Project Management Body of Knowledge (PMBOK, 2017) gave the ten project management knowledge areas essential for the smooth delivery of any project (construction inclusive). These knowledge areas are considered as critical operations/process needed for effective project delivery. These key knowledge areas are project integration management, project scope management, project schedule management, project cost management, project quality management, project resource management, project communications management, project risk management, project procurement management, and project stakeholder management. It is believed in the context of this study that digitalising these processes can lead to the attainment of better service delivery through digitalisation. Thus, construction organisations seeking to attain digital transformation must strive to possess reasonable maturity in these mentioned process areas as shown in Table 8.3.

Strategy

There is no naysaying that the strategy adopted by an organisation to have more satisfied clients and make a better profit is the result of the business model being adopted, as a synergy between business models and organisation's strategy is evident in past studies (Osterwalder, 2004; Richardson, 2008; Wikström *et al.*, 2010). A company directly or indirectly adopts a business model that defines how the company hopes to create, deliver, and retain value (Teece, 2010). This is deemed necessary as it defines how the company will ensure customers get value for their money, and at the same time it shows how the company intends to make its customers pay for the value they are receiving and converting these payments into profit for the organisation. Bouwman *et al.* (2018) described a business model as strategies targeted at ensuring that value is created for both the customer and the business. Zott and Amit (2008) defined business models as "*the structure, content, and governance of transaction between the focal firm and its exchange partners such as customers, vendors, collaborators*". A further description shows that a business model is a system which offers the fitting together of pieces of a business (Magretta, 2002) which allows business managers to adequately understand and develop their organisation in a more holistic manner (Pekuri *et al.*, 2015). In a nutshell, business models show organisation's management/business owners' understanding of the desire of their customers, how these customers want to be serviced, and the best approach the organisation can take to ensure these desires are met to get paid and make profit while keeping the customers satisfied (Teece, 2010). However, Bouwman *et al.* (2018) noted that to gain better competitive advantage, the business model adopted by an organisation requires constant modification that will result in significant and observable changes in the services delivered to their customers. This modification was described as innovation introduced into the organisation's business model.

In construction, the traditional approach to business is to acquire new projects and ensure the successful delivery of such a project in line with the requirements stated from the project's onset (Pekuri *et al.*, 2015). The value of this new project is solely based on the client's needs as to what the end product should be, and to a large extent, this is defined by the project design (Bertelsen & Koskela, 2004). However, Osterwalder (2004) stated that for most of the world's best organisations, their business is built around achieving value that is defined by both the organisation and the customer. Thus, Pekuri *et al.* (2015) hinted that in construction organisations, "value" should be a crucial issue long before the commencement of a project. This "value" should not be in terms of the end product but how the organisations hope to help their clients achieve projects that serve their intended purposes. To achieve this, following the need for an innovative modification in an organisation's business model as observed by Bouwman *et al.* (2018), digital thinking right from every project's early stage is necessary.

Ernest and Young Global (2018) stated that for organisations aiming to gain a better competitive advantage over their counterparts, it is imperative to instil innovative digital concepts in all aspects of their strategic approach. It was further noted that while adoption of new digital technologies can help organise company's activities, integrate systems, make communication more efficient, recognise efficiencies, increase production, reduce cost and more, it is highly unrealistic to attain these benefits without a well-planned strategy. It is also important to note that with the advent of ICTs, customers now have access to adequate information and solutions to their needs at low cost (Teece, 2010). This means that customers now have more choices; the supplies of which are more transparent. Thus, aside from acquiring technologies, having the right strategy to provide valuable services to these customers and retain them in the process, using these technologies is important. In affirmation, Kane *et al.* (2015) submitted that the strength of digitalisation is not in the individual technologies being adopted but rather lies in the company's ability to integrate them into their daily activities or business models to transform their businesses and operations. It was further noted that the distinction between digital leaders from non-digital leaders is a clear digital strategy that is mixed with a culture and leadership aimed at driving the digital transformation.

In the context of this study, this strategy dimension which is evident in the three major areas (process, position, and path) of the DCT is deemed crucial as it deals with the strategies being adopted by construction organisations towards attracting and retaining construction clients, improving their service delivery, and making an optimum profit. Gill *et al.* (2016) named this dimension as the "insight" dimension, which deals with how well a company uses customer and business data to measure success and inform strategy. Newman (2017) sees this dimension as the strategic management of the organisation brand, the ecosystem in which the organisation functions, and the stakeholders. It also involves the finance and investment, market and customer, portfolio, idea and innovation. De Carolis *et al.* (2017) named this dimension the monitoring and control dimension, which involves how the processes are monitored and controlled through the evaluation of feedback received from their execution. In construction, these feedbacks are obtained from stakeholders of projects. Bennis (2013) has noted the need for organisations to adopt strategies that encourage information-driven transparency derived through digital technologies.

Similarly, Day (2011) and Quinton *et al.* (2018) observed that there is the need for organisations to imbibe the culture of forecasting customers' future needs and at the same time envisage their competitors' move. Also, organisations can adopt the strategy of adaptive experimentation. This involves investing in smaller digital technology projects with a view to gain reasonable insights to the adoption of digital tools and at the same time understand the risk involved in their usage. These experiments will give the required result at a lower risk level and in a safer manner. McKeown and Philip (2003) have also identified this issue of risk by stating that establishing risk (digital risk) as a cultural norm within organisations is a crucial strategy for digital technology adoption and digital transformation. Matt *et al.* (2015) submitted that organisations that desire to be digitally transformed need to formulate a strategy that will cut across other business strategies and allows management to properly coordinate business activities, identify priorities, implement, and at the same time oversee transformations that occur because of new technologies.

Research and development (R&D) also plays an essential role in an organisation's ability to sense opportunities lurking within their domain. In this case, construction organisations need to promote meaningful R&D that will help unearth the necessary digital technologies required for digital transformation and competitive advantage of the organisation. Other vital variables such as decision-making, digital technology selection, avoiding decision error, project and service boundary, and choice of market selection are evident in the DCT as major sub-attributes relating to the strategy (Teece, 2007). Based on the preceding, the 14 sub-attributes to measure the strategy capability of a construction organisation is listed in Table 8.4.

Table 8.4 Strategy dimension of DCMM

Strategy	Authors
Research and development	Teece (2007)
Digital technology selection	Teece (2007)
Forecasting client's need	Day (2011), Quinton *et al.* (2018)
Choice of market selection	Teece (2007)
Project and service boundary	Teece (2007)
Client's feedback evaluation	Bharadwaj *et al.* (2013), De Carolis *et al.* (2017)
Investment in digital technology	Coltman (2015), Dyk and Schutte (2012), Newman (2017)
Decision-making procedure	Teece (2007)
Digitalised performance measurement	Teece (2007)
Error-proofing approaches	Teece (2007)
Information and communication management	Bennis (2013)
Digital alignment of business	Newman (2017), Valdez-de-Leon (2016)
Digital risk culture within the organisation	McKeown and Philip (2003)
Open innovation within the organisation	Teece (2007)

Gaps in existing models

Following the extraction of the four major dimensions discussed earlier, it was noticed that there exist some gaps in the body of the literature regarding the dimensions needed for digital transformation, particularly in the construction industry. Considering the absence of a DCMM designed purposely for construction organisations, these observed gaps were introduced to give a holistic maturity model suitable for assessing digital capability maturity of construction organisations. While these gaps might have been mentioned in some earlier assessed models, they were in most cases mentioned as sub-attributes. These identified gaps are digital partnering and environment, and they are discussed in relation to the digital transformation of construction organisations.

Digital partnering

The construction business exists within a highly competitive and dynamic environment. The advent of different digital technologies that promise better service delivery does not help construction organisations as they are required to possess dynamic capabilities that will match that of the dynamic digital environment. Thus, The Economist Intelligence Unit (EIU, 2015) stated that organisations, in general, gain a competitive advantage in the ever-competitive digital environment, partnering with other organisations with the same vision is vital. It was stated that organisations are fast recognising that it is tougher to be alone; thus, they are "*entering into partnerships of one form or another to develop digital capabilities*".

Vollmer and Egor (2014) have earlier noted that the first rule towards successful digital transformation is never to innovate alone. Going through the journey of digital transformation with like-minded partners is essential. Therefore, based on this notion, it is important to first understand the concept of partnering and the critical factors needed to achieve successful digital partnering. By doing so, we can give proper guidelines on how construction organisations can adopt partnering successfully to attain digital transformation.

To understand the concept of digital partnering, exploring the meaning of partnering is essential. In literature, the coming together of two or more organisations to achieve a common goal has been described using different terms such as network, partnership, partnering, integration, strategic alliance, etc. (Cheng *et al.*, 2000). Nevertheless, some differences exist between these different terms when adopted, particularly between partnership and partnering, which are often used interchangeably. According to Perez (2020), the distinct difference between both is that partnership is a legal entity or a type of business, while partnering is a way of running a business. The concept of partnering in construction became popular due to the poor project performance experienced within the construction industry (Alderman & Ivory, 2007; Cheng *et al.*, 2000; Cook & Hancher, 1990). In partnering, two or more entities come together to make a long-term commitment with the sole aim of fulfilling some set objectives using each other's resources (Construction Industry Institute, 1991). Nunez *et al.* (2018) submitted that partnering is a management tool employed to ensure that all parties to a contract are committed to a contract's set goal.

In clearly defining the components of partnering, Nystrom adopted the Wittgenstein's method of definition which argued that *"there are complex networks of overlapping similarities among the things that fall under a complex concept"* (Nystrom, 2005). Through this idea, components with overlapping similarities were identified, and a partnering flower was generated. The partnering flower was first presented in 2005 and was found highly useful by researchers. Yeung *et al.* (2007) adopted it in defining what an alliance contract is. However, some shortcomings were observed in the initial version of the partnering flower designed, which led to the redevelopment of a more holistic design that is considered useful for this current study.

Nystrom's partnering flower theory

To unearth the components of partnering that need to be in place to achieve partnering goals, Nystrom (2005) applied the Wittgenstein's approach by looking at the frequency of mention of different components of partnering within the body of knowledge. The partnering flower shows two important components at the centre with seven other less important components acting as the petals surrounding these two major components. These two components are "trust" and "mutual understanding/common goal". They are surrounded by relationship-building activities, predetermined dispute resolution method, economic incentive contracts, facilitator, openness, continuous and structured meetings, and choosing working partners. It is interesting to note that the adoption of the key components of the partnering flower in a partnering contract varies from contract to contract. While the two key centre components (trust and mutual understanding) remain common, the addition of some of the remaining components varies. The additional component necessary for one particular partnering project might not be needed for another.

Digital partnering for construction digitalisation

Considering the low level of adoption of digital technologies among construction organisations, digital partnering can be a viable option for attaining digital transformation. It is important to note that while partnering can either be one-off or strategic, if construction organisations are to attain digital transformation through partnering, a long-term goal must be set. This implies that while it is good to partner with other organisations on a common goal of achieving a project using digital technologies, it is also imperative to create a long-term relationship that will enable continuous development through continuous partnership on the subsequent project. Following the generic definition of partnering, "Digital Partnering" is defined as "*a long-term commitment between two or more organisations to achieve digital transformation by maximising the effectiveness of each participant's digital resources*" (Aghimien *et al.*, 2020).

It is important to note that digital partnering in the context of this book holds a different meaning to "digital collaboration", which has been described as the interaction between industry and the digital world (Oke *et al.*, 2018). While "digital collaboration" is the use of digital technologies (such as online meetings and webinars, team chatrooms, shared spreadsheets, etc.) to connect a broader network of participants together, "digital partnering" in this context is the coming together of two or more organisations to share digital resources to have a better competitive advantage and achieve digital transformation in the process.

Vollmer and Egor (2014) observed that partnering with other organisations is a critical strategy for businesses seeking to grow in unfamiliar markets, tap into new customer segments, or sell additional products or services. This situation applies to the digital transformation of construction organisations. Through effective digital partnering, a considerable level of digital capability can be achieved. Bharadwaj *et al.* (2013) also saw the need to establish a partnership with external actors and manage relations to attain proper digital business strategy. It was also stated that for digital transformation to occur, organisations must be able to design, structure, and manage networks that will afford them significant capabilities that are complementary to what is existing within the organisation. Bennis (2013), Boström and Celik (2017), and Kane *et al.* (2015) have earlier noted that employees' digital skills can be upgraded through a proper alliance with external bodies.

Out of all the models reviewed earlier in Chapter 6, only Luftman's (2000) strategic alignment maturity assessment model explicitly stated a form of the alliance as a standalone dimension. The dimension was named partnership and entails the business perception of IT value, IT's role in strategic business planning, shared goals, risk, rewards/penalties, IT program management, relationship/trust style, and business sponsor. However, this model aimed to assess business alignment maturity with IT and does not give a clear view of what these selected components entail or where they emanated from. Similarly, Boström and Celik's (2017) maturity model of digital strategising discussed partnering under their ecosystem dimension, which was described as a modification of Luftman's partnership dimension. However, just like every other model that has mentioned the need for an alliance (IDC, 2015; Valdez-de-Leon, 2016), this model treated partnership as only a variable. This variable was merged with the ability to react fast to ecosystem changes and sensing environmental changes and responding to new IT initiatives. Hence the name "ecosystem".

Research conducted by EIU (2015) discovered that most organisations' executives believe that for organisations to leverage technology in the future, they will have to be part of a network. This network can be seen as a form of digital partnering, which requires organisations to be open to new and innovative approaches to conducting their businesses. This process can be overwhelming as EIU (2017) stated that "the challenges of choosing the right partner, building successful partnerships and managing the risks inherent with adding a third party to one's technology ecosystem can be formidable". Therefore, the onus is on the organisation's management to carefully think and plan the whole digital partnering process and all stages of the organisation's operations. This can be achieved by taking into consideration the critical success factors for effective partnering as it applies to digital partnering and the organisation's operations (Nystrom, 2005; Cheng & Li, 2004).

EIU (2017) noted that in achieving digital capabilities required for digital transformation, organisations must first determine if they are willing to build these capabilities alone or buy them. By buying, organisations make a conscious decision of acquiring needed capabilities (especially technology) from a third party rather than building one themselves. Sometimes, developing digital capabilities from outside can prove more advantageous and even economical than building from scratch within the organisation (Day, 2011). Even though these capabilities are obtained from outside, still the decision on how to deploy these capabilities and attain successful transformation lies with the organisation. After deciding to acquire digital capabilities from outside the organisation, the next step is to ensure a successful partnership. Organisations need to clearly understand the critical success factors for digital partnering and ensure they are put in place. This implies that if digital transformation is to be attained among construction organisations through effective partnering, understanding the key components necessary for digital partnering is important.

Drawing from the general critical success factors for partnering discussed in 19 construction partnering studies, the significant factors related to digital partnering are discussed here. It is imperative to note that while these factors are deemed crucial for digital partnering to deliver projects through the use of digital technologies and to attain digital transformation of partnering organisations in the process, other factors not mentioned explicitly may have some underlining occurrence within those mentioned. Similarly, as Nystrom (2005) observed, not all the identified factors that may exist within a digital partnering relationship were created. Considering the importance of trust and mutual understanding, other factors might be added to create a successful process. However, these factors must be agreed upon right from forming the partnering agreement and made known clearly to all parties involved.

Choosing the right digital partner – Because digitalisation has been a slow process among construction organisations, and even a non-starter for some (Aghimien *et al.*, 2018; Osunsanmi *et al.*, 2018), choosing the right partner will be the first significant step (after deciding to partner) towards attaining successful digital partnering and achieving digital transformation. It has earlier been observed that choosing the right working partner and building a reliable team are very important to attain successful partnering as partnering is said to be largely dependent on good personal interaction (Cheng & Li, 2002; Conley & Gregory, 1999; Kadefors, 2004; Nystrom, 2005). In terms of digital partnering, construction organisations must be cautious in selecting organisations to partner with. This is because digital partnering aims to attain digital transformation of the organisation in the long run while delivering better services in the process using digital technologies. Considering the scarceness of resources within the construction industry, especially in the area of technology, knowledge, information, and personnel as observed by Chan *et al.* (2004), choosing organisations with complementary resources to what is obtainable within the organisation is necessary. Construction organisations with lesser digital capabilities, especially in terms of resources, will do well to partner with organisations that have a better advantage in these areas within and outside the construction industry.

Trust in digital partner – The importance of trust in partnering has been reiterated over and over again in past studies. It is crucial for the effective management of relationships between project partners (Das & Teng, 2001; Wong & Cheung, 2005). Building trust among digital partners can go a long way in mitigating relationship risks as through trust, partners can become closer, and settling issues that may arise in the future can be easier. This is necessary as even with the best plans in place, a partnering process can still collapse if the trust element is taken out of the process (EIU, 2017). Chan *et al.* (2004) submitted that trust in partnering could emanate in each partner's form, believing that the other party is living up to expectation in the partnering relationship. This trust can be in the aspect of partners' competence, integrity, or just intuition (Hartman, 2003). Digital partners can, therefore, create a trustworthy environment whereby there is belief in the partners' digital competence and their integrity as it relates to fulfilling the partnering agreements.

Mutual understanding/common goal – The sole desire of every partnering process is to attain a common goal while achieving individual objectives in the process. Some studies have described this process of achieving a common goal to attain a win-win outcome (Cook & Hancher, 1990; Kadefors, 2004). However, Nystrom (2005) noted that this win-win scenario for partners as related to project outcome is in most cases, unattainable. As a result, a mutual understanding between parties in reaching a certain level of compromise is necessary for a long-term functioning partnering relationship to exist between parties. Therefore, if construction organisations are to attain digital transformation through an effective partnering system, a mutual understanding to achieve a common goal and stay consistent with the partnering relationship's objectives is necessary. This can be made clear right from the onset of the partnering process through effective communication.

Top management support – The role of management in the adoption of new ideas cannot be overemphasised. Cheng *et al.* (2000) noted that any organisation's management is responsible for the formulation of business strategy and direction. This implies that the desire to partner or not to partner depends on the top management's decision and what they consider to be best for the organisation. Similarly, EIU (2017) has stated that in the quest to attain digital transformation through the development of organisation's digital capability, the management of the organisation must decide whether they will like to develop such capability by themselves or rely on the help of a third party. This is a pointer to the fact that the decision to partner lies with the top management. Also, top management is responsible for the partnering relationship's success, which they initiate through a proper commitment to the course.

Positive conflict resolution and openness – Since all partners have their objectives which they try to achieve in the quest to attain the common goal set for the partnering process, there is the tendency of conflicts to arise. Having a predefined positive way towards addressing such conflicts when they occur is necessary. A positive conflict resolution system can be achieved through a win-win solution (Chan *et al.*, 2004; Cook & Hancher, 1990; Nystrom, 2005). This positive open communication can stem from openness among partners. Without openness in terms of information sharing and having an effective communication system in place, the partnering process will not survive. Through openness, partners' weaknesses and strengths are understood, and their challenges are revealed. Each partner knows where the other is lagging and knows how to compliment them to achieve the partnering relationship's set goal. Similarly, trust can be built when partners are open to one another as they reveal the challenges they face, believing that their partners can help come up with positive solutions. Bayramoglu (2001) has earlier stated that with openness and effective communication system, problems can be identified early, and measures can be put in place to cushion their effect should they occur eventually. Considering the dynamic digital world, where new digital technologies are being developed by the day and updation of existing ones is endless, digital partners must be more open in sharing ideas with one another in order to survive.

Structured digital workshops and training – Digital partnering relationship is built to deliver effective projects through digital capabilities of partners and help partners improve their digital capability and attain digital transformation. Therefore, there is a need for continuous and structured meetings wherein problems arising during projects are assessed, solutions are developed, and ideas needed for improvement are shared (Nystrom, 2005). In the same vein, training and workshops on digital technologies can be organised to keep partners and their staff updated on technological happenings and at the same time improve their knowledge in terms of the use of these ever-increasing technologies.

Long-term commitment – The journey towards digital transformation is a continuous process which will require organisations to keep improving and maintaining their digital capabilities. Thus, a one-off partnering relationship might not be the best for organisations with a long-term digital transformation goal. Therefore, it is advised that construction organisations consider the path of strategic partnering, which, according to Cheng and Li (2002), seeks long-term relationships between partners and the attainment of strategic goals. Digital partners should be ready to stick together in the face of unanticipated problems (Bresnen & Marshal, 2000) and focus on achieving digital transformation that is attained through an increase in their digital capabilities developed by an effective partnering system.

Efficient coordination – The digital partnering process requires efficient coordination by a facilitator with considerable experience and understanding in managing partnering relationships. Hellard (1996) and Nystrom (2005) noted that a facilitator's job is to ensure relevant issues are discussed during workshops and that partners can put forth their view and that their perspectives are respected. With digital partnering, wherein most organisations might not have full understanding of the technology needed for certain activities or have the necessary resources, there is the need for someone to facilitate the relationship between partners in a way that no party feels left out or feel they are giving too much to the system.

Based on the aforementioned, eight major factors were extracted for the measurement of digital partnering for construction digitalisation, and they are discussed in Table 8.5.

Table 8.5 Digital partnering dimension of DCMM

Digital partnering	*Authors*
Digital partner selection	Cheng and Li (2002), Conley and Gregory (1999), Kadefors (2004), Nystrom (2005)
Long-term commitment to digital partnering	Black *et al.* (2000), Cheng *et al.* (2000), Chan *et al.* (2004), Cheng and Li (2002)
Top management support	Cheng *et al.* (2000), EIU (2017)
Conflict resolution in digital partnering	Chan *et al.* (2004), Cook & Hancher (1990), Nystrom (2005)
Trust in digital partners	Chan *et al.* (2004), Das and Teng (2001), EIU (2017), Wong and Cheung (2005).
Common goal in partnering process	Black *et al.* (2000), Conley and Gregory (1999), Kadefors (2004), Nystrom (2005), Packham *et al.* (2003); Tang *et al.* (2009)
Digital partnering coordination	Chan *et al.* (2004), Cheng and Li (2002), Cheng *et al.* (2000), Conley and Gregory (1999), Kadefors (2004)
Structured digital workshop and training	Nystrom (2005)

Environment

In the quest to adopt new technology and innovations within organisations, the external environment within which such an organisation operates plays a crucial role. This is evident in the "path" dimension of the DCT, which shows that the future of an organisation is impacted either positively or negatively by the opportunities and threats within the environment (Teece *et al.*, 1997). Studies tailored towards understanding the factors influencing the adoption and intention to adopt technologies and innovations have emphatically stressed the need to understand the external environment (Chatterjee *et al.*, 2002; Govender & Pretorius, 2015; Hsu *et al.*, 2006; Quinton *et al.*, 2018; Teo *et al.*, 2003; Tidd, 2001; Tsuja & Marinō, 2013). In light of this, several adoption models and framework developed have placed emphasis on the organisation's external environment. An example is the Technology-Organisation-Environment (T-O-E) framework, which explicitly assesses the influence of environment on technology adoption (Tornatzky & Fleischer, 1990). Several studies have adopted this framework in assessing technology and innovation adoption within organisations (Awa *et al.*, 2017; Gutierrez *et al.*, 2015). Another framework worth mentioning is the Decision-maker-Technology-Organisation-Environment developed by Thong (1999), which is an advancement of the T-O-E with the inclusion of the decision-making aspect. This framework also considers the function of the organisation's external environment in the adoption of technology.

Thus, if construction organisations are to be digitally matured, there must be considerable consciousness regarding the external variables that could drive their digitalisation attainment or impede it. It has been noted that being cautious of external business environment issues and responding promptly to them are important for the survival of most organisations as many businesses have failed as a result of not paying attention to this crucial aspect (Moysés *et al.*, 2010; Zhang *et al.*, 2011). McGee and Sawyerr (2003) described external environment as those physical and social variables that can impact the decision-making of an organisation but exist outside the confines of the organisation. Tsuja and Marinō (2013) went further to describe external environment as those external variables that are complex and uncertain and can cause a change in the reflection of an organisation. This environment is dynamic as it is characterised by high speed and frequency of change within a short time (Robbins & Coulter, 2005). Relating this to the construction industry, which has been characterised as dynamic and complex, it is evident that construction organisations need to be cautious of their external environment in the quest for digital transformation.

Several theories, such as the systems theory and the contingency theory, have been postulated to help decipher the key components of an organisation's external environment. However, this current study borrows from the knowledge of "institutional theory" (DiMaggio & Powell, 1983) in understanding the components of an external environment. The institutional theory sees the organisation as an institution which operates within an institutional environment, and it categorises the different mechanisms through which institutional changes might occur based on the kind of pressure exacted by the institutional environment on the organisation. Roberts and Greenwood (1997) have earlier noted that organisations tend to be under diverse institutional pressures such as social expectations and cultural norms that may conflict with efficiency in the quest for efficiency in service delivery. Thus, this theory was deemed useful for this current study because understanding the maturity level of construction organisations in handling external pressures will help proffer possible measures by which these organisations can overcome or embrace these pressures and attain digital transformation in the process. Past studies have also adopted principles of this theory in understanding how environmental issues force organisations to adopt certain concepts or innovations (Awa *et al.*, 2017; Govender & Pretorius, 2015; Gutierrez *et al.*, 2015; Hsu *et al.*, 2006; Quinton *et al.*, 2018; Teo *et al.*, 2003).

The institutional theory

Aldrich (1979) stated that one factor that organisations cannot ignore is other organisations. In view of this, DiMaggio and Powell (1983) posited that organisations do not compete for only customers and needed resources, but they also compete for "*power and institutional legitimacy, for social as well as economic fitness*". As these organisations compete, institutional theory believes that they come under pressure to conform to the shared notions of appropriate forms and behaviours. This is because non-conformance may lead to questioning the organisation's legitimacy and this might affect the organisation's ability to secure resources and social support (DiMaggio & Powell, 1983; Teo *et al.*, 2003). Thus, DiMaggio and Powell (1983) concluded that institutional theory is a useful tool for understanding most organisations' key issues.

Studies have continued to promote the institutional theory as a crucial viewpoint in understanding the role of environment in organisation's adoption of innovation (Liu *et al.*, 2010; Rogers *et al.*, 2007; Teo *et al.*, 2003). This promotion is as a result of the fact that the environment wherein an organisation operates exacts considerable amount of pressure on the organisation, and this pressure when transferred through operational channels can influence the organisation's intention to adopt innovations (Liu *et al.*, 2010; Roberts & Greenwood, 1997). It is believed that the institutional environment dictates the rules in terms of social expectations and norms necessary for organisations to have the right structure and behaviour as well as the right operations and practices (DiMaggio & Powell, 1983; Scott, 1995). Thus, in deciding whether to adopt innovations or not, organisations will first consider the institutional expectations and norms. Through this consideration, an evaluation of the costs to be incurred and the benefits to be derived from such adoption are assessed. Based on the results of these considerations and assessments, the organisation can, therefore, position itself properly in order not to be taken aback by uncertainties (DiMaggio & Powell, 1983; Liu *et al.*, 2010; Scott, 1995).

Based on the aforementioned, it is clear that organisations are subject to the pressure of being isomorphic with their institutional environment (Teo *et al.*,

2003). DiMaggio and Powell (1983) identified three major types of isomorphic pressures from the institutional environment. These are coercive pressure which emanates from political influences and the problem of legitimacy; mimetic pressure which results from a standard response to uncertainty; and normative pressure which is associated with professionalism.

The environment in construction digitalisation

In the quest to develop a digitalisation capability maturity model for construction organisations, the influence of external construction environment on the digitalisation of construction organisations cannot be overlooked. Using the institutional theory, it is obvious that construction organisations face significant pressure from the environment. Their maturity in handling these pressures and converting them into a positive drive to achieve digitalisation is important to note. Therefore, this study assessed the external environment from the three pressures of the institutional theory – coercive, mimetic, and normative pressure.

Coercive pressure – This results from "*both formal and informal pressures exerted on organisations by other organisations upon which they are dependent and by cultural expectations in the society within which organisations function*" (DiMaggio & Powell, 1983). Coercive pressure may emanate from other organisations with higher resource power, regulatory bodies, and even parent corporations. This is because a body or an organisation with high resource power tends to control scarce and important resources. Such an organisation may require those organisations that depend on it for resources to adopt structures or programmes that serve their interests. In a bid to survive, these organisations with lesser resource power may be forced to adhere to these requirements. Similarly, if an organisation decides to operate in a manner or direction that is opposite of what is attainable in its industry, such an organisation runs the risk of properly maintaining such operations and maintaining a good relationship with others within the industry. This implies that organisations will have to conform with the pattern within their industry for their operations to be workable. They tend to conform with formal policies, models, and programmes of the industry to survive. Thus, it can be said that coercive pressure, in simple terms, comes from a place of dependence (Teo *et al.*, 2003).

In construction, different types of relationships and interdependence exist. Contractor–supplier relationships, contractor–client relationships, contractor–consultant relationships are a few types of relationship that show some level of interdependency between each party (Wang, 2000). This means that a construction organisation, operating as a contractor or consultant, is bound to experience some sort of pressure from its trading partners. If the digital transformation is to be attained within such an organisation, there must be a significant maturity level in handling this pressure. Assessing the ability of construction organisations to handle suppliers' and clients' pressure in terms of using relevant digital technologies is important in determining their digitalisation maturity level.

Previous studies have also deemed this dimension crucial to technology and innovation adoption (Gupta *et al.*, 2013; Gutierrez *et al.*, 2015; Hsu *et al.*, 2006;

Low *et al.*, 2011; Quinton *et al.*, 2018). Similarly, it has been stated that the government plays a significant role in the construction industry's operations. They make certain legislation that significantly affects the industry. Therefore, construction organisations aiming to attain digital transformation must be ready to maximise some of these legislations in the adoption of digital technologies. Therefore, it is crucial to assess construction organisations' maturity in handling government legislation relating to the use of digital technologies.

Mimetic pressure – Over time, an organisation may copy ideas and activities from other organisations within their environment and become like those they have copied from (DiMaggio Powell, 1983). A key driver of this imitation is uncertainty (Quinton *et al.*, 2018). This can result from a poor understanding of required technology, ambiguous goals, or symbolic uncertainty within the environment (DiMaggio Powell, 1983). It is believed that an organisation will want to emulate the footsteps of other organisations that share the same goals, produce the same commodity, have similar suppliers and customers, and face similar challenges with them (Teo *et al.*, 2003). Awa *et al.* (2017) submitted that due to the unstable nature of the business environment, every organisation would continue to monitor others within the same industry and somehow mimic their activities to stay competitive. This type of pressure causes retaliatory and endless vicious circle as organisations mimic others, those they are mimicking do something else to stay ahead of the competition, and this circle continues (Awa *et al.*, 2015, 2017). However, in mimicking others, organisations will always pick only positive, innovative traits of their successful competitors (Oliveira & Martins, 2011; Gibbs & Kraemer, 2004; Pang & Jang, 2008).

There is no naysaying that the construction industry is highly competitive. If construction organisations survive and remain competitive within the industry, they must seek that unique and sustainable edge over their competitors. However, in doing this, construction organisations face the pressure of mimicking their competitors and becoming like them. This is not a wrong concept as studies have revealed that this type of pressure can lead to competitive advantage, considerable improvement in the use of technologies, and even development of innovative ideas based on what they have seen from their competitors. Therefore, construction organisations gunning for digitalisation of its services must view competitive pressure in a positive light and use this competitive intensity as a drive towards adopting relevant digital technologies and even introducing other digital technologies into the system that are not originally being used within the industry.

Normative pressure – This arises from professionalisation (DiMaggio & Powell, 1983). Studies have shown that when an organisation has ties with other organisations that have adopted the use of certain technologies or innovations, it becomes easier for them to learn about such technologies and innovations, their associated cost and the benefit to be derived thereof. By doing so, they are encouraged to adopt these technologies and innovations (Teo *et al.*, 2003). Through this, a norm is created within the industry where these organisations exist. Powell and DiMaggio (1991) noted that through these norms, consensus is achieved within the industry, increasing these norms and their potential influence on organisational behaviour. Quinton *et al.* (2018) described this normative pressure as the *"expectations associated with professionalisation"*. It was further stated that this type of pressure could stem from existing rules and regulations that an organisation needs to comply with to remain in business within an industry or to have social legitimisation. Hsu *et al.* (2006) have earlier noted that these rules and regulations are in most cases driven by trade associations or the broader regulatory environment. Previous studies have also indicated that this type of pressure may emanate from firm–supplier and firm–customer relations, professional, trade, business, and other key organisations (Powell & DiMaggio, 1991; Teo *et al.*, 2003).

Evidently, construction organisations do not exist in isolation. The construction industry, as a whole, plays a significant role in how organisations deliver their services. Several norms have been created over time. An example is the use of certain technologies in place of humans to provide safety during construction. Going against these norms might earn such organisations severe sanction and even loss of legitimacy. Thus, it can be said that to conform to the norms of the industry, construction organisations are put under certain pressures. Handling these pressures specifically in relation to rules and regulations regarding the use of digital technologies, ensuring safety and security (of persons and information) while using digital technologies, and complying to the environmental impact factor of the use of such digital technologies are essential to the digital transformation of construction organisations. In the same vein, it is important to understand the industry market scope in order to adopt relevant digital technologies (Alshamaila *et al.*, 2013; Awa *et al.*, 2017; Govender & Pretorius, 2015; Hsu *et al.*, 2006).

Based on the aforementioned, Table 8.6 shows the external environment sub-attributes that can influence the digital transformation of construction organisations and for which maturity needs to be attained for construction organisations to attain digitalisation.

Table 8.6 Environment dimension of DCMM

Environment	Authors
Digital technology suppliers' pressure	Gupta *et al.* (2013), Gutierrez *et al.* (2015), Hsu *et al.* (2006), Low *et al.* (2011), Quinton *et al.* (2018).
Environmental impact assessment	Awa *et al.* (2017)
Competitors' influence	Awa *et al.* (2017), Gutierrez *et al.* (2015), Hsu *et al.* (2006), Low *et al.* (2011), Quinton *et al.* (2018).
Government legislation	Brender and Markov (2013), Govender and Pretorius (2015), Hsu *et al.* (2006)
Industry rules and regulations	Alshamaila *et al.* (2013), Awa *et al.* (2017), Govender and Pretorius (2015), Hsu *et al.* (2006)
Clients' pressure	Gupta *et al.* (2013), Gutierrez *et al.* (2015), Hsu *et al.* (2006), Low *et al.* (2011), Quinton *et al.* (2018).
Market scope	Alshamaila *et al.* (2013), Awa *et al.* (2017)

Specification and justification of the structural model

This study aims to develop a DCMM for construction organisations. The model is geared towards creating a means for construction organisations to assess their current digital capability and improve on areas of shortcomings. As stated in Chapter 7, the model is conceptualised upon the classification of the DCT developed by Teece *et al.* (1997), while the maturity level in the Capability Maturity Model (CMM) developed at the Software Engineering Institute at Carnegie Mellon University was adopted for measuring the maturity level for each capability dimension of the model. The DCT noted that for a business to survive and gain sustainable competitive advantage, the business must possess the capability to sense opportunities and threats within the business environment. It must also have the capability to seize these opportunities and at the same time reconfigure or transform the structure and resources of the organisation to fit with the seized opportunities (Schoemaker *et al.*, 2018).

Further to this, Teece and Pisano (1994) have earlier mentioned that these capabilities can exist in three major areas of the organisation's operations. These are the organisational and managerial process, shaped by the organisation's assets position and the evolutionary path the organisation has adopted. Since this current book is geared towards developing a model for the construction industry, categorising the different capability dimensions needed for digitalisation maturity into these three different areas becomes obvious.

Based on the review of existing models from diverse industries other than construction, four major capability areas were discovered and categorised based on the organisation's process, position, and path. These four capability dimensions are process capability that relates to the organisational and managerial process; technology and people's capabilities, which have to do with the organisation's asset position, and strategies capability that cuts across the process, position, and the evolutionary path of the organisation. Similarly, with the Nystrom partnering flower theory (Nystrom, 2005), the need for a digital partnering capability became obvious, especially for construction organisations struggling to fully grasp what digitalisation is and what lies ahead of them in the journey towards digital transformation. This became the first gap in existing models that were introduced to the conceptualised model. This dimension lies in the "position" capability of the DCT as external relations is a specific asset that determines the position of an organisation. Lastly, the institutional theory by DiMaggio and Powell (1983) was explored to understand how pressure from the environment wherein these construction organisations operate can influence their digitalisation. Based on the theory, the second gap in existing models was discovered, and this was termed "environment". This gap lies in the "path" domain of the DCT where the organisation's future depends on the path it has adopted and opportunities within the immediate environment.

Based on this knowledge, the conceptualised DCMM in this study is rooted in the three DCT pillars of process, position, and path. These capability dimensions are technology, process, people, strategy, digital partnering, and environment.

Furthermore, the literature review to reveal the outcome variables of digitalisation was conducted to understand the inherent benefits construction organisations stand to gain from being digitalised. These outcome variables are both subjective and objective. This study combines both types of outcome variables on the notion that sometimes objective variables might be misleading, equally subjective upon further scrutiny, and in most cases insufficient in the development of policies (Aigbavboa, 2013).

Defining the maturity levels

In terms of the maturity level for the DCMM, as earlier stated, this current study draws from the CMM traditional five levels of maturity (see Figure 8.1).

Maturity level one – "initiating level"

According to Paulk *et al.* (1993), the CMM described organisations at the first level as the initial level wherein organisations do not have a stable and friendly environment for developing and maintaining software. Organisations lack the right management practices, and the benefits of being derived from good software engineering practices are not being enjoyed due to ineffective planning. Valdez-de-Leon (2016) described this phase as "not started" in its six-level maturity model, meaning organisations at this stage are yet to take steps towards digital transformation. This stage was called the "zero levels" as nothing has commenced. In construction, it is most unlikely to see an organisation presently without a single form of a digital trace either in the design or construction of projects. Hence, the zero levels might not apply to this current study. Therefore, the first maturity level for this study is defined as the "Initiating level". Newman (2017) described organisations at this phase to be at the early discussion stage and beginning to incorporate digital tools into their businesses. According to Valdez-de-Leon (2016), organisations at the initiating level have decided to move towards digital business and take steps towards digital transformation. Gill *et al.* (2016) described this level as the sceptic level where organisations are just beginning their digital journey and are unsure of the outcome.

For this current study, organisations at this level are those that have the vision to be digitally transformed and discussion on how to go about it is in place. However, they lack the right capabilities needed towards achieving this feat, they tend to be sceptical of what the outcome of digitalisation will be, and as a result, they do not enjoy the inherent benefits of digitalisation due to ineffective planning. Organisations at this level must strive to sense and seize opportunities to gain improved capability in the six different digitalisation dimensions identified to move to the second level.

Maturity level two – "repeatable level"

The CMM described the second level as the "repeatable" level where policies for managing a software project and procedures to implement those policies are

Maturity Levels

Initiating Level
Traces of digital technology usage can be found within the construction organisation. These technologies are used to make service delivery easier and not necessarily to attain digitalisation of the organisation. The construction organisation at this level is sceptical about the outcome of taking the digitalisation route.

Repeated Level
There is an understanding of the concept of digitalisation, but the capability needed for its attainment is fuzzy to the organisation. The organisation learns from repeated actions. The organisation is willing to learn from the digitalisation process as it unfolds and is willing to use knowledge gained from past implementation on new ones to fully grasp the whole digital transformation concept and process.

Defined Level
The capabilities needed for digitalisation is known. There are clear objectives of attaining digital transformation and plans towards achieving these objectives are put in place.

Managed Level
The construction organisation has quantifiable goals geared towards achieving digitalisation. Emphasis is on the productivity of the different capability dimensions, and this is measured as part of the organisation's measurement programme. Performance measurement at this level is consistent.

Optimising Level
The construction organisation here is a digital disruptor of the construction industry. The organisation continuously improves its digital capability, promotes innovation and regularly leads construction industry's discussion on digital transformation.

Digitalisation Capability Dimension

- Technology
- People
- Process
- Strategy
- Digital Partnering
- Environment

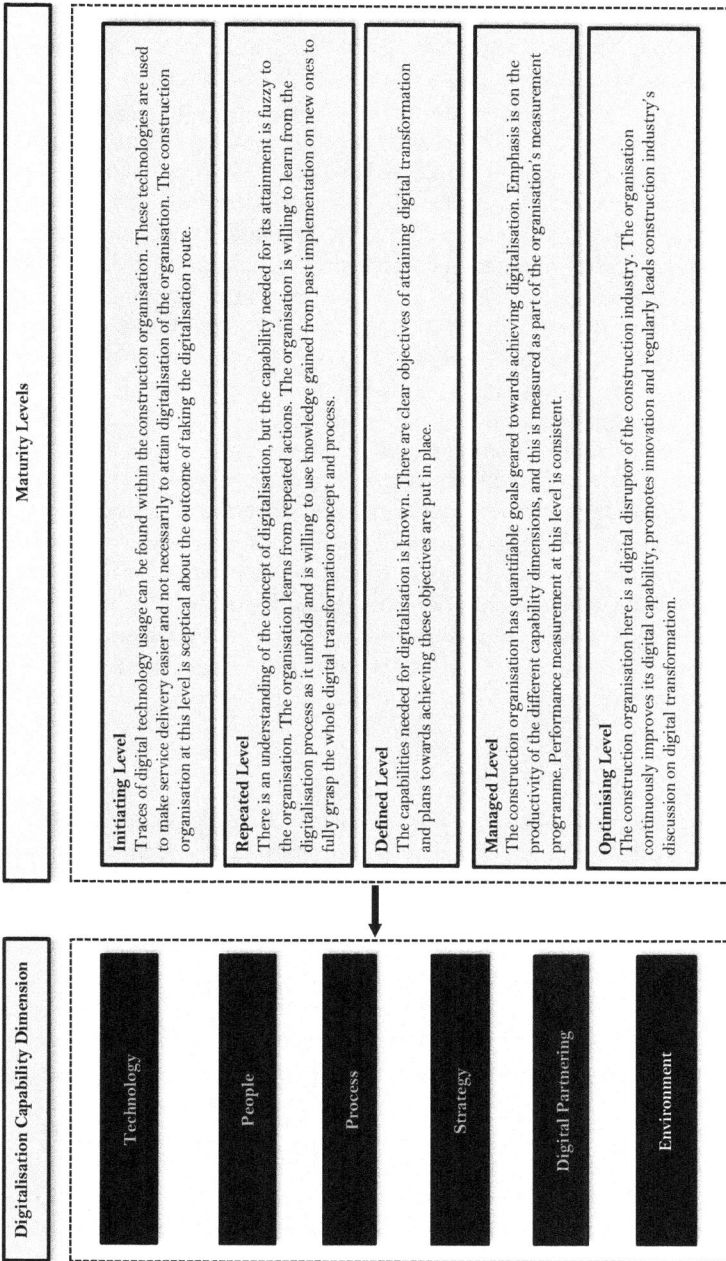

Figure 8.1 Digitalisation capability dimensions and the maturity levels adopted

Source: Author's compilation (2020)

established. At this level, experience harnessed from similar projects handled in the past is used to plan and manage new projects, hence making it a repeatable level (Paulk *et al.*, 1993). In the modified model (i.e., CMMI) this level was described as "managed", which involves some planned project management processes. The process here is also often reactive, and the necessary process discipline is in place to repeat earlier successes on projects with similar applications (Carnegie Mellon SEI, 2005). According to Macchi and Fumagalli (2013), the processes here are partially planned, and performance analysis is mostly dependent on individual practitioner's experience and competences. Luftman (2000) further described this level as the "committed" level whereby management within an organisation is committed to the new line of action and are willing to build upon past knowledge gained. For Valdez-de-Leon (2016), this second level is the "enabling" level where organisations implement initiatives that will form the foundation of their digital business. Newman (2017) described this level as the "emerging" level whereby discussions are advanced and beginning to be incorporated into all of organisation's daily operations.

From the aforementioned, different names abound for this level, but the concept remains the same. For this current study, the maturity level two is seen as the repeatable level whereby organisations at this level have clear plans for digitalisation and have put policies and procedures in place to see these plans come to life. These organisations are willing to learn from the digitalisation process as it unfolds and is ready to use the knowledge gained from past implementation on new ones to grasp the whole digital transformation concept and process fully. Improvement in the policies and procedures and willingness to learn and implement the knowledge gained from previous projects will see construction organisations at this level move to the third level.

Maturity level three – "defined level"

The CMM described organisations at the third level as the defined level where the standard process for developing and maintaining software across the organisation is documented. This documentation includes both the software's engineering and management processes (Paulk *et al.*, 1993). These processes are then combined into a single entity. In their development of a maturity model for IT management, Becker *et al.* (2009) described this level as a defined level where IT performance measurement process is defined and relies on consolidated IT reporting implemented with desktop tools. In the CMMI, this level is also named the defined level where both management and engineering activities are documented, standardised, and integrated into a standard process for the organisation. This process is largely proactive (Carnegie Mellon SEI, 2005). According to Macchi and Fumagalli (2013) at this level, the process is planned, and semi-quantitative analyses are done periodically to define good practices and management procedures. Gill *et al.* (2016) described organisations at this level as "Adopters" because they tend to invest in digital skills and infrastructures. Valdez-de-Leon (2016) described this level as the "integrating" level wherein the organisation's initiative is integrated across the organisation to support end-to-end capabilities. In the same vein, Newman (2017) described this stage as the "performing" stage whereby the organisation has set clear objectives and formulated a plan that is being followed throughout the company.

Similarly, for this current study, the maturity level three is seen as the defined level for construction organisations aiming for digital transformation. The organisations at this stage have figured out the standard process and capabilities needed for the digitalisation of their services and at the same time have set clear objectives and formulated a plan that is being followed throughout the organisation.

Maturity level four – "managed level"

Level 4 of the CMM is seen as the "Managed" level where organisations set quantitative goals for their products and processes. At this level, productivity is measured as part of the organisation's measurement programme. Software processes are instrumented with well-defined and consistent measurements (Paulk *et al.*, 1993). A similar adoption was made by Becker *et al.* (2009). They described organisations at this level as those with a formalised and automated IT performance measurement process supported by a powerful IT performance dashboard. The CMMI described this level as "quantitatively managed" level, which involves the detailed measurement of the software process and product quality. At this level, the software process and products are quantitatively understood and controlled (Carnegie Mellon SEI, 2005). In the same vein, Macchi and Fumagalli (2013) described this level as the quantitatively managed level where process performance is measured and allows special variations to be detected. In line with the CMM, Luftman (2000) noted that this stage could be seen as the "improved" or "managed" level. Also, Newman (2017) named this level, the "advancing" level wherein the organisation expands on its plan and objectives to develop new and innovative ideas to advance its capabilities in this area. Valdez-de-Leon (2016) ascribed the name "optimising" level to the fourth maturity level since, at this stage, organisations can fine-tune their digital initiatives within the different dimensions to further increase performance.

This current study adopts the CMM nomenclature of "managed" level for organisations at the fourth maturity level of digitalisation. Construction organisation at this level possesses set reasonable quantitative goals to be met in the quest for digitalisation of their processes. Construction organisations at this level place emphasis on the productivity of their different capability dimensions and these are measured as part of the organisation's measurement programme. The performance measurement is consistent and allows for variability between the set goals and the outcome to be seen and corrected. Through performance measurement, these organisations can expand their plan and objectives to develop new and innovative ideas to advance their digitalisation capabilities.

Maturity level five – "optimising level"

The last level of maturity is the "optimising" level, whereby the entire organisation is focused on continuous process improvement. The means to identify weaknesses and strengthen the process proactively, with the goal of preventing the occurrence of defects is available within organisations at this level (Paulk *et al.*, 1993). According to Becker *et al.* (2009), the organisation possesses cross-linked IT performance measurement process that is implemented with company-wide integrated business intelligence tools. The CMMI noted that at this level, continuous process improvement is enabled by quantitative feedback from the process and from piloting innovative ideas and technologies (Carnegie Mellon SEI, 2005). The sole aim at this stage is to ensure continuous improvement of what has been implemented (Macchi & Fumagalli, 2013). In the view of Valdez-de-Leon (2016), organisations operating at this level can be said to be at the "pioneering" level as they often break new grounds and advance the state of digital practice within their industry. On the other hand, Newman (2017) described this level as the "leading" level where organisations are considered thought leaders in the area of digital transformation, regularly leads industry discussions on the topic, and has mastered the subject area.

Considering the dynamic nature of both the construction industry and the digital world (Navon, 2005; Solis, 2016), it is imperative for organisations that have been able to understand the capabilities needed for digitalisation and have been able to set objectives in motion towards being digitally transformed, to improve on their current positions continuously. To this end, the fifth maturity level for the digitalisation of construction organisations is described as the "optimising" level. Organisations at this level should possess the capability to be digital disruptors of the construction industry.

Outcomes of successful digitalisation

The use of digital technologies such as BIM tends to help deliver construction projects in time and within budget. This is because, with the use of BIM, clashes in designs are detected early in the project and rework is therefore avoided. This saves the time wasted on the rework of construction works and its associated cost (Hashim *et al.*, 2013). In the same vein, the use of electronic tendering and procurement platforms saves time and money for construction clients and contractors (Berger, 2016). Furthermore, it has been observed that with the use of robotics and automation, the need for an employee to be doing repetitive works is reduced, thus, reducing overall project cost to some extent (Kamaruddin *et al.*, 2016). Digital technologies can also help deliver more quality projects that the clients and other project stakeholders can be satisfied with. For instance, it has been observed that the use of robotics and automation can bring about increased productivity and significant satisfaction to the client and other stakeholders as a more durable and precise construction project can be achieved. Also, BIM can help deliver value for construction clients' money, thus leading to increased client satisfaction (Hashim *et al.*, 2013; Kamaruddin *et al.*, 2016; Vaduva-Sahhanoglu *et al.*, 2016).

The use of autonomous robots can increase the social sustainability of construction projects in terms of health and safety as risky jobs are done using robots (Oke *et al.*, 2017; Ruggiero *et al.*, 2016). Hager *et al.* (2016) and Sakin and Kiroglu

(2017) submitted that with the use of technologies such as 3D printers, hazardous works are done by the printer leaving the less hazardous works to be carried out by humans. By doing so, human lives are preserved. Using wearable devices with embedded sensors can help monitor workers' health issues and detect problems early. This is done through measuring and capturing health data of workers. Early detection of issues around workers overworking, stress, and subsequent absenteeism allow measures to be put in place before they affect the overall project outcome (Salento, 2017; Hudson, 2017).

Through digitalisation, effective project monitoring of construction projects and plant and equipment can be achieved (Bogue, 2018). More jobs can be created, and new industry can spring forth (Fonseca, 2018; Pîrjan & Petroşanu, 2013), thus improving the economy through job creation. Also, through the adoption of technologies, construction organisations can gain a competitive advantage over their counterparts (Bharadwaj *et al.*, 2013; Kamaruddin *et al.*, 2016). To gain competitive advantage, organisations also need to be innovative (Bharadwaj *et al.*, 2013). Past studies have revealed that construction industries have lagged in implementing technological innovations to its full potential, and, therefore, have been seen to lack innovativeness. Thus, through adopting applicable and available digital technologies, construction industries and organisations can be more innovative in their service delivery. According to Labonnote *et al.* (2016), the use of digital technologies such as 3D printing can also lead to increased innovative ideas, especially in terms of construction designs. The concept provides an avenue for considerable design freedom innovativeness. Based on the preceding the expected outcomes of the successful construction digitalisation can be summarised as follows:

- Improved digital culture
- Improved digital uptake
- Improved digital readiness
- Digitally transformed construction organisation
- Better cybersecurity of an organisation's data
- Effective data management in construction projects
- Better project delivery to cost
- Better project delivery to time
- Better project conformance to quality
- Increased competitiveness in the global market
- Increased competitiveness in the local market
- Increased innovativeness
- Creation of digital employment opportunities
- Increased productivity
- Increased client satisfaction
- Better social sustainability in a project (H&S)
- Effective procurement system
- Effective project monitoring and control
- Better overall project performance

Theorising the conceptualised digitalisation capability maturity model

The study theorised that a construction organisation that will attain significant digital transformation, provide better service delivery for its client, and in the process obtain a competitive edge over its competitors must consider its maturity in six dynamic digital capabilities, namely technology, people, process, strategy, digital partnering and environment (see Figure 8.2). Through these dynamic digitalisation capabilities, managers within these organisations can effectively utilise both the internal and external digital competencies of the organisation to produce new business models and better service delivery.

Figure 8.3 gives an overview of the conceptualised DCMM needed for the digital transformation of construction organisations, especially in developing countries where construction digitalisation is still at its infancy stage. Construction organisations will have to attain improved maturity in each of the identified capability areas. These organisations can assess their maturity level on a scale of one to five for each of the itemised sub-attributes under each capability area. A cumulative of

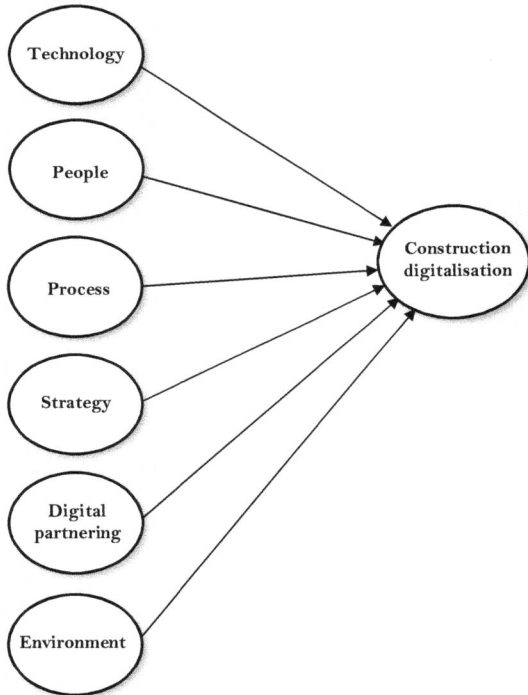

Figure 8.2 Conceptual digitalisation capability model

Initiating level (1) ⟶ Optimising level (5)

Technology
- Technology planning
- Emerging technologies optimisation
- Big data analytics
- Additive manufacturing
- Cloud computing/Connected things (IoT, IoE, IoS)
- Standard IT systems
- Design simulation and virtualisation
- Robotics and automation in construction
- Digital media marketing
- Cybersecurity
- Technology governance
- Continuous technology upgrades
- Flexible technology usage
- The synergy of technology with other organisation's resources

People
- Digital technical know-how of personnel
- Digital knowledge management by the organisation
- Reskilling of workforce
- Continuous learning of personnel
- Top management support
- Digital culture within an organisation
- Organisation's positive change attitude
- Digital empowerment of personnel
- Personnel's innovativeness
- Attracting digital talent
- Retaining digital talent

Process
- Digitalised planning process
- Automated construction process
- Digitalised schedule management
- Digitalised resource management
- Management of integrated service
- Value delivery and quality management
- Digitalised stakeholder management
- Digitalised scope management
- Digitalised cost management process
- Digitalised procurement process
- Digital risk management
- Change management
- Communication

Strategy
- Research and development
- Digital technology selection
- Forecasting client's need
- Choice of market selection
- Project and service boundary
- Client's feedback evaluation
- Investment in digital technology
- Decision-making procedure
- Digitalised performance measurement
- Error-proofing approaches
- Information and communication management
- Digital alignment of business
- Digital risk culture within the organisation
- Open innovation within the organisation

Digital partnering
- Digital partner selection
- Long-term commitment to digital partnering
- Top management support
- Conflict resolution in digital partnering
- Trust in digital partners
- Common goal in the partnering process
- Digital partnering coordination
- Structured digital workshop and training

Environment
- Clients' pressure
- Digital technology suppliers' pressure
- Competitors' influence
- Government legislation
- Industry rules and regulations
- Market scope
- Environmental impact assessment

Process   Position   Path

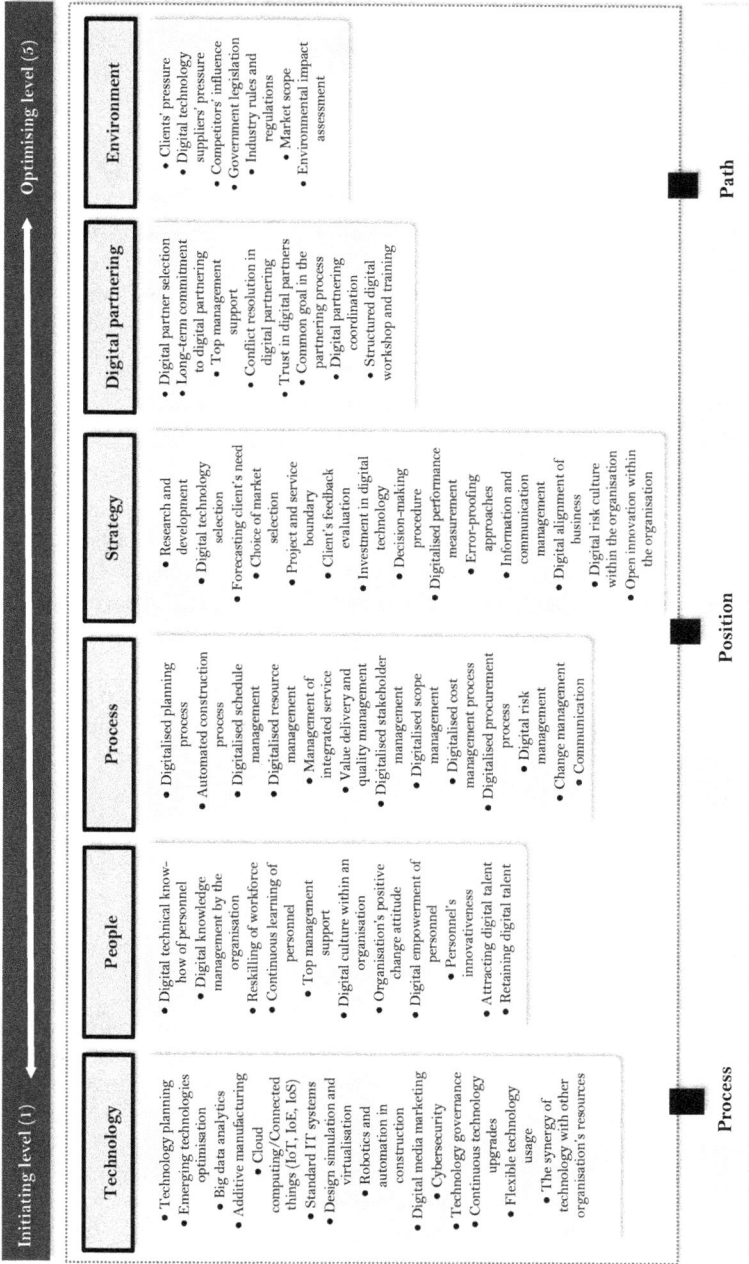

Figure 8.3 Conceptualised DCMM (dimensions and sub-attributes)

the attained scale for each sub-attribute can be aggregated to show the maturity level under each capability dimension. The cumulative score attained for each capability dimension can also be aggregated to determine the overall digitalisation capability of the organisation. For instance, taking the technology dimension, a typical question when assessing the organisation's digital capability can follow: On a scale of one to five with one being "very low" and five being "very high", what is the level of technology planning within your organisation? Or, on a scale of one to five with one being "very low" and five being "very high", to what extent are emerging technologies such as IoT, Big Data, AI, AR, VR, Robotics and automation, machine learning, 3D printing among others adopted and optimised within your organisation? The selected scale for each question denotes the organisation's maturity level with respect to that particular variable. In the end, the average of the aggregated cumulative will determine the maturity level of the organisation within the dimension.

Summary

This chapter discussed the different dimensions and sub-attributes of a conceptualised DCMM for construction organisations. Drawing from existing maturity models and the gaps therein, the chapter theorised that construction organisations seeking to be digitally transformed must assess their capability maturity in six major dimensions: technology, people, process, strategy, digital partnering, and environment. These organisations must endeavour to increase their maturity in each dimension areas at a five-level maturity scale. An optimised level of maturity must be targeted to gain digital transformation, provide better service delivery for its client, and obtain a competitive edge over its competitors. The next chapter explores the suitability and applicability of the conceptualised DCMM in a developing country's construction industry.

References

Agarwal, R., Chandrasekaran, S. and Sridhar, M. (2016). *Imagining Construction's Digital Future*. Capital Project and Infrastructure, McKinsey and Company. Available at: www. mckinsey.com/industries/capital-projects-and-infrastructure/our insights/imagining-con structions-digital-future [accessed 19–11–2020]

Aghimien, D.O., Aigbavboa, C.O. and Oke, A.E. (2018). Digitalisation for effective construction project delivery in South Africa. *Contemporary Construction Conference: Dynamic and Innovative Built Environment* (CCC2018), Coventry, 5–6 July, pp. 3–10

Aghimien, D.O., Aigbavboa, C.O. and Oke, A.E. (2019). Digitalisation of construction organisations in South Africa: A dynamic capability theory approach (2019). Advances in ICT in design, construction and management in architecture, engineering, construction and operations (AECO). *Proceedings of the 36th CIB W78 2019 Conference*, Northumbria University, Newcastle, United Kingdom, 18–20 September, pp. 74–83

Aghimien, D.O., Aigbavboa, C.O. and Oke, A.E. (2020). Critical success factors for digital partnering in construction organisations – a Delphi study. *Engineering, Construction and Architectural Management*, 27(10): 3171–3188

Aigbavboa, C. (2013). *An Integrated Beneficiary Centred Satisfaction Model for Publicly Funded Housing Schemes in South Africa*. A PhD thesis submitted to the Post Graduate School of Engineering Management, University of Johannesburg, Johannesburg

Alderman, N. and Ivory, C. (2007). Partnering in major contracts: Paradox and metaphor. *International Journal of Project Management*, 25: 386–393

Aldrich, H. (1979). *Organizations and Environments*. Prentice-Hall, Englewood Cliffs, NJ.

Alshamaila, Y., Papagiannidis, S. and Li, F. (2013). Cloud computing adoption by SMEs in the north east of England: A multi-perspective framework. *Journal of Enterprise Information Management*, 26(3): 250–275

Andriole, S. (2016). 15 must-have technology capabilities for digital transformation (The final Scream). Available at: www.forbes.com/sites/steveandriole/2016/09/20/15-must-have-technology-capabilities-for-digital-transformation-the-final-scream/#3677e4a2fe15

Astebro, T. (2002). Noncapital investment costs and the adoption of CAD and CNC in U.S. Metalworking industries. *The RAND Journal of Economics*, 33: 672–688

Awa, H., Baridam, D. and Nwibere, B. (2015). Demographic determinants of e-commerce adoption: A twist by location factors. *Journal of Enterprise Information Management*, 28(3): 325–346

Awa, H.O., Ojiabo, O.O. and Orokor, L.E. (2017). Integrated technology-organization-environment (T-O-E) taxonomies for technology adoption. *Journal of Enterprise Information Management*, 30(6): 893–921

Bayramoglu, S. (2001). Partnering in construction: Improvement through integration and collaboration. *Leadership and Management in Engineering*, 1(3): 39–43

Becker, J., Knackstedt, R. and Pöppelbuß, J. (2009). Developing maturity models for IT management. *Business and Information Systems Engineering*, 1(3): 213–222

Bennis, W. (2013). Leadership in a digital world: Embracing transparency and adaptive capacity. *MIS Quarterly*, 37(2): 635–636

Berger, R. (2016). Digitization in the construction industry. *Think Act*: 1–16

Bertelsen, S. and Koskela, L. (2004). Construction beyond lean: A new understanding of construction management. *Proceedings of the International Group Lean Conference*. Helsingør, Denmark, p. 12

Bharadwaj, A., El Sawy, O., Pavlou, P. and Venkatraman, N. (2013). Digital business strategy: Toward a next generation of insights. *MIS Quarterly*, 37(2): 471–482

Bhaskaran, S. (2013). Structured case studies: Information communication technology adoption by small-to-medium food enterprises. *British Food Journal*, 115: 425–447

Bibby, L. and Dehe, B. (2018). Defining and assessing industry 4.0 maturity levels – case of the defence sector. *Production Planning and Control*, 29(12): 1030–1043

Black, C., Akintoye, A. and Fitzgerald, E. (2000). An analysis of success factors and benefits of partnering in construction. *The International Journal of Project Management*, 18(6): 423–432

Bogue, R. (2018). What are the prospects for robots in the construction industry? *Industrial Robot*, 45(1): 1–6

Borges, M., Hoppen, N. and Luce, F.B. (2009). Information technology impact on market orientation in e-business. *Journal of Business Research*, 62: 883–890

Boström, E. and Celik, O.C. (2017). *Towards a Maturity Model for Digital Strategizing – A Qualitative Study of How an Organization Can Analyse and Assess Their Digital Business Strategy*. IT Management Master Thesis submitted to the Department of informatics, UMEA Universitet, Sweden

Bouwman, H., Nikou, S., Molina-Castillo, F.J. and de Reuver, M. (2018). The impact of digitalization on business models. *Digital Policy, Regulation and Governance*, 20(2): 105–124

Brender, N. and Markov, I. (2013). Risk perception and risk management in cloud computing: Results from a case study of Swiss companies. *International Journal of Information Management*, 33(5): 726–733

Bresnen, M. and Marshall, N. (2000). Motivation, commitment and the use of incentives in partnerships and alliances. *Journal Construction Management and Economics*, 18(5): 587–598

Carnegie Mellon SEI. (2005). *Capability Maturity Model® Integration (CMMI) Overview*. Carnegie Mellon University, Pittsburgh, PA

Castagnino, S., Rothballer, C. and Gerbert, P. (2016). *What's the Future of the Construction Industry?* World Economic Forum. Available at: www.weforum.org/agenda/2016/04/building-in-the-fourth-industrial-revolution/ [accessed 19–11–2020]

Chan, A.P.C., Chan., D.W.M., Chiang, Y.H., Tang, B.S., Chan, E.H.W. and Ho, K.S.K. (2004). Exploring critical success factors for partnering in construction projects. *Journal of Construction Engineering and Management*, 130(2): 188–198

Chatterjee, D., Grewal, R. and Sambamurthy, V. (2002). Shaping up for e-commerce: Institutional enablers of the organisational assimilation of web technologies. *MIS Quarterly*, 26: 65–89

Cheng, E. and Li, H. (2002). Construction partnering process and associated critical success factors: Quantitative investigation. *Journal of Management in Engineering*, 18(4): 194–202

Cheng, E.W.L. and Li, H. (2004). Development of a practical model of partnering for construction projects. *Journal of Construction Engineering & Management*, 130(6): 790–798

Cheng, E.W.L., Li, H. and Love, P.E.D. (2000). Establishment of critical success factors for construction partnering. *Journal of Management in Engineering*, 6(2): 84–92

Coltman, T., Tallon, P., Sharma, R. and Queiroz, M. (2015). Strategic IT alignment: Twenty-five years on. *Journal of Information Technology*, 30(2): 91–100

Conley, M.A. and Gregory, R.A. (1999). Partnering on small construction projects. *Journal of Construction Engineering & Management*, 125(5): 320–324

Construction Industry Institute. (1991). *In Search of Partnering Excellence*. Construction Industry Institute, Austin, TX, pp. 17–11

Cook, E.L. and Hancher, D.E. (1990). Partnering: Contracting for the future. *Journal of Management in Engineering*, 6(4): 431–446

Das, T.K. and Teng, B.S. (2001). Trust, control and risk in strategic alliances: An integrated framework. *Organization Studies*, 22(4): 253–285

Day, G.S. (2011). Closing the marketing capabilities gap. *Journal of Marketing*, 75: 183–195

De Carolis, A., Macchi, M., Negri, E. and Terzi, S. (2017). A maturity model for assessing the digital readiness of manufacturing companies. In Lödding, H., *et al.* (Eds.), *APMS 2017. IFIP Advances in Information and Communication Technology. (IFIPAICT, Vol 513)*. Springer Verlag, Berlin, pp. 13–20

Delgado, J.M.D., Oyedele, L., Akinade, O., Bilal, M., Akanbi, L. and Ajayi, A. (2017). BIM data model requirements for structural monitoring of built assets. *Proceedings of Environmental Design and Management International Conference (EDMIC)*, Held Between 22–24 of May, at the Obafemi Awolowo University, Ile-Ife, Nigeria, pp. 118–126

DiMaggio, P. and Powell, W. (1983). The iron cage revisited: Institutional isomorphism and collective rationality in organizational fields. *American Sociological Review*, 48(2): 147–160

Dyk, L.V. and Schutte, C.S.L. (2012). Development of a maturity model for telemedicine. *South African Journal of Industrial Engineering*, 23(2): 61–72

The Economist Intelligence Unit. (2015). *Connecting Companies Strategic Partnerships for The Digital Age*. A Telstra Reports. Available at: http://connectedfuture.economist.com/

connecting-capabilities/article/connecting-companies-strategic-partnerships-for-the-digital-age/ [accessed 6–8–2020]

The Economist Intelligence Unit. (2017). *Digital Refinement: C-Level Executives Hone Their Transformation Skills.* Choosing the Right Technology Partner. Available at: https://eiu-perspectives.economist.com/sites/default/files/Choosing_the_right_technology_partner.pdf [accessed 6–8–2020]

Ehlhardt, H. (2014). Computer aided cost estimating. *International Conference on Engineering and Product Design Education*, University of Twente, The Netherlands, 4–5 September

Eidhoff, A.T., Stief, S.E., Voeth, M. and Gundlach, S. (2016). Drivers of digital product innovation in firms: An empirical study of technological, organizational, and environmental factors. *International Journal of Economics and Management Engineering*, 10(6): 1888–1892

Ernest and Young Global. (2018). How are engineering and construction companies adapting digital to their businesses? Available at: www.ey.com/Publication/vwLUAssets/EY-Digital-survey/$File/EY-Digital-survey.pdf

Fonseca, L.M. (2018). Industry 4.0 and the digital society: Concepts, dimensions and envisioned benefits. *Proceedings of the International Conference on Business Excellence*, 34: 386–397

Gatignon, H. and Xuereb, J.M. (1997). Strategic orientation of the firm and new product performance. *Journal of Marketing Research*, 34(1): 77–90

Gebhardt, G.F., Carpenter, G.S. and Sherry, J.F. (2006). Creating a market orientation: A longitudinal, multiform, grounded analysis of cultural transformation. *Journal of Marketing*, 70: 37–55

Gerbert, P., Castagnino, S., Rothballer, C., Renz, A. and Filitz, R. (2016). *Digital Engineering and Construction: The Transformative Power of Building Information Modelling.* The Boston Consulting Group, Boston, MA

Gibbs, L. and Kraemer, K. (2004). A cross-country investigation of the determinants of scope of e-commerce use: An institutional approach. *Electronic Markets*, 14(2): 124–137

Gill, M., VanBoskirk, S., Evans, P.F., Nail, J., Causey, A. and Glazer, L. (2016). The digital maturity model 4.0. Benchmarks: Digital business transformation playbook. Available at: www.forrester.com

Goodrum, P.M. and Haas, C.T. (2004). Long-term impact of equipment technology on labour productivity in the U.S. Construction industry at the activity level. *Journal of Construction Engineering and Management*, 130: 124–133

Govender, N.M. and Pretorius, M. (2015). A critical analysis of information and communications technology adoption: The strategy-as-practice perspective. *Acta Commercii*, 15(1): 1–13

Grant, K., Edgar, D., Sukumar, A. and Meyer, M. (2014). 'Risky business': Perceptions of e-business risk by UK small and medium sized enterprises (SMEs). *International Journal of Information Management*, 34: 99–122

Gupta, P., Seetharaman, A. and Raj, J. (2013). The usage and adoption of cloud computing by small and medium businesses. *International Journal of Information Management*, 3(5): 861–874

Gutierrez, A., Boukrami, A. and Lumsden, R. (2015). Technological, organisational and environmental factors influencing managers' decision to adopt cloud computing in the UK. *Journal of Enterprise Information Management*, 28(6): 788–807

Hager, I., Golonka, A. and Putanowicz, R. (2016). 3D printing of buildings and building components as the future of sustainable construction? *Procedia Engineering*, 151: 292–299

Harrigan, P., Ramsey, E. and Ibbotson, P. (2011). Critical factors underpinning the e-CRM activities of SMEs. *Journal of Marketing Management*, 27: 503–529

Hartman, F.T. (2003). *Ten Commandments of Better Contracting: A Practical Guide to Adding Value to an Enterprise Through More Effective SMART Contracting*. ASCE Press, Reston, VA, pp. 235–260

Hashim, N., Said, I. and Idris, N.H. (2013). Exploring E-procurement value for construction companies in Malaysia. *Procedia Technology*, 9: 836–845

Hellard, R.B. (1996). The partnering philosophy – a procurement strategy for satisfaction through a team work solution to project quality. *Journal of Construction Procurement*, 2(1): 41–55

Hsu, P.F., Kraemer, K.L. and Dunkle, D. (2006). Determinants of e-business use in US firms. *International Journal of Electronic Commerce*, 10(4): 9–45

Hudson, V. (2017). The digital future of the infrastructure industry: Innovation 2050. *Balfour Beatty*. Available at: www.balfourbeatty.com/how-we-work/public-policy/innovation-2050-a-digital-future-for-the-infrastructure-industry/ [accessed 19–12–2019]

Ibem, E.O. and Laryea, S. (2016). E-tendering in the South African construction industry. *Journal of Construction Management*, 17(4): 310–328

International Data Corporation (IDC). (2015). Digital transformation: Benchmarking assessment IDC recommendations. Available at: www.idc.com

Jones, R., Souranta, M. and Rowley, J. (2013). Entrepreneurial marketing: A comparative story. *The Service Industries Journal*, 33: 705–719

Kadefors, A. (2004). Trust in project relationships – inside the black box. *International Journal of Project Management*, 22(3): 175–182

Kamaruddin, S., Mohammed, M.F. and Mahbub, E. (2016). Barriers and impact of mechanisation and automation in construction to achieve better quality products, Malaysia. *Procedia – Social and Behavioural Sciences*, 222: 111–120

Kane, G.C., Palmer, D., Phillips, A.N., Kiron, D. and Buckley, N. (2015). *Strategy, Not Technology, Drives Digital Transformation*. MIT Sloan Management Review and Deloitte University Press, Westlake, TX

Kenny, J. (2006). Strategy and the learning organization: A maturity model for the formation of strategy. *The Learning Organization*, 13(4): 353–368

Khin, S. and Ho, T.C.F. (2018). Digital technology, digital capability and organizational performance: A mediating role of digital innovation. *International Journal of Innovation Science*, 11(2): 177–195

Labonnote, N., Rønnquist, A., Manum, B. and Rüther, P. (2016). Additive construction: State of the art, challenges and opportunities. *Automation in Construction*, 72(3): 347–366

Liu, H., Ke, W., Wei, K.K., Gu, J. and Chen, H. (2010). The role of institutional pressures and organisational culture in the firm's intention to adopt internet-enabled supply chain management systems. *Journal of Operations Management*, 28: 372–384

Low, C., Chen, Y. and Wu, M. (2011). Understanding the determinants of cloud computing adoption. *Industrial Management and Data Systems*, 111(7): 1006–1023

Luftman, J. (2000). Assessing business-IT alignment maturity. *Strategies for Information Technology Governance*, 4(14): 1–50

Macchi, M. and Fumagalli, L. (2013). A maintenance maturity assessment method for the manufacturing industry. *Journal of Quality in Maintenance Engineering*, 19(3): 295–315

Magretta, J. (2002). Why business models matter. *Harvard Business Review*, 80(5): 86–92

Matt, C., Hess, T. and Benlian, A. (2015). Digital transformation strategies. *Business & Information Systems Engineering*, 57(5): 339–343

McGee, J.E. and Sawyerr, O.O. (2003). Uncertainty and information search activities: A study of owner – managers of small high-technology manufacturing firms. *Journal of Small Business Management*, 41: 385–401

McKeown, I. and Philip, G. (2003). Business transformation, information technology and competitive strategies: Learning to fly. *International Journal of Information Management*, 23(1): 3–24

McLaughlin, S.A. (2017). Dynamic capabilities: Taking an emerging technology perspective. *International Journal of Manufacturing Technology and Management*, 31(1–3): 62–81

Moysés, J.F., Kestelman, H.N., Beecker, L.C., Jr. and Torres, M.C.S. (2010). *Strategic Planning and Management in Healthcare Organizations*. Publishing Company FGV, Rio de Janeiro

Navon, R. (2005). Automated project performance control of construction projects. *Automation in Construction*, 14: 467–476

Newman, M. (2017). Digital maturity model (DMM): A blueprint for digital transformation (TM Forum White Paper). Available at: www.tmforum.org

Nunez, N.P., Del Puerto, C.L. and Jeong, D. (2018). Development of a partnering maturity assessment tool for transportation agencies. *The Journal of Legal Affairs and Dispute Resolution in Engineering and Construction*, 10(4): 1–12

Nystrom, J. (2005). The definition of partnering as a Wittgenstein family resemblance concept. *Construction Management and Economics*, 23: 473–481

Oke, A.E., Aghimien, D.O., Aigbavboa, C.O. and Koloko, N. (2018). Challenges of digital collaboration in The South African construction industry. *Proceedings of the International Conference on Industrial Engineering and Operations Management*, Bandung, Indonesia, March 6–8, pp. 2472–2482

Oke, A.E., Aigbvaboa, C.O. and Mabena, S. (2017). Effects of automation on construction industry performance. Second international conference on mechanics, materials and structural engineering (ICMMSE 2017). *Advances in Engineering Research (AER)*, 102: 370–374

Oliveira, T. and Martins, M. (2011). Literature review of information technology adoption models at firm level. *The Electronic Journal Information Systems Evaluation*, 14(1): 110–121

Osterwalder, A. (2004). *The Business Model Ontology – A Proposition in Design Science Approach*. PhD dissertation, Department of Business and Economics, University of Lausanne, Switzerland

Osunsanmi, T.O., Aigbavboa, C.O. and Oke, A.E. (2018). Construction 4.0: The future of South Africa construction industry. *World Academy of Science, Engineering and Technology International Journal of Civil and Environmental Engineering*, 12(3): 206–212

Packham, G., Thomas, B. and Miller, C. (2003). Partnering in the house building sector: A subcontractor's view. *International Journal of Project Management*, 21(5): 327–332

Pang, M. and Jang, W. (2008). Determinants of the adoption of ERP within the T-O-E framework: Taiwan's communications industry. *Journal of Computer Information Systems*: 94–102

Paulk, M., Curtis, B., Chrissis, M. and Weber, C. (1993). Capability maturity model for software, Version 1.1. Available at: http://wwwsei.cmu.edu/pub/documents/93.reports/pdf/tR24.93.pdf

Pekuri, A., Pekuri, L. and Haapasalo, H. (2015). Business models and project selection in construction companies. *Construction Innovation*, 15(2): 180–197

Peltier, J.W., Schibrowsky, J.A. and Zhao, Y. (2009). Understanding the antecedents to the adoption of CRM technology by small retailers: Entrepreneurs vs owner-managers. *International Small Business Journal*, 27: 307–336

Peltier, J.W., Zhao, Y. and Schibrowsky, J.A. (2012). Technology adoption by small businesses: An exploratory study of the interrelationships of owner and environmental factors. *International Small Business Journal*, 30: 406–431

Perez, D. (2020). The difference between partnership & partnering. Available at: https://smallbusiness.chron.com/difference-between-partnership-partnering-38333.html [accessed 7–6–2020]

Pîrjan, A. and Petroşanu, D.M. (2013). The impact of 3D printing technology on the society and economy. *Journal of Information Systems & Operations Management*, 7(2): 163–173

Powell, W.W. and DiMaggio, P.J. (1991). *The New Institutionalism in Organisational Analysis.* University of Chicago Press, Chicago

Project Management Body of Knowledge PMBOK. (2017). *A Guide to Project Management Body of Knowledge*, 6th edition. Project Management Institute, Philadelphia, PA

Quinton, S., Canhoto, A., Molinillo, S., Pera, R. and Budhathoki, T. (2018). Conceptualising a digital orientation: Antecedents of supporting SME performance in the digital economy. *Journal of Strategic Marketing*, 26(50): 427–439

Richardson, J. (2008). The business model: An integrative framework for strategy execution. *Strategic Change*, 17(5/6): 133–144

Robbins, S.P. and Coulter, M. (2005). *Management*, 8th edition. Prentice Hall, Saddle River, NJ

Roberts, P.W. and Greenwood, R. (1997). Integrating transaction cost and institutional theories: Toward a constrained-efficiency framework. *Academy of Management Journal*, 22(2): 346–373

Rogers, K.W., Purdy, L., Safayeni, F. and Duimering, P.R. (2007). A supplier development program: Rational process or institutional image construction? *Journal of Operations Management*, 25(2): 556–572

Ruggiero, A., Salvo, S. and St. Laurent, C. (2016). *Robotics in Construction.* Worcester Polytechnic Institute, Worcester, MA

Sakin, M. and Kiroglu, Y.C. (2017). 3D printing of buildings: Construction of the sustainable houses of the future by BIM. *Energy Procedia*, 134: 702–711

Salento, A. (2017). Digitalisation and the regulation of work: Theoretical issues and normative challenges. *Artificial Intelligence and Society Journal*, 33: 369–378

Schoemaker, P.J.H., Heaton, S. and Teece, D.J. (2018). Innovation, dynamic capabilities, and leadership. *California Management Review*, 6(1): 15–42

Scott, R.W. (1995). *Institutions and Organizations.* Sage, Thousand Oaks, CA

Sepasgozar, S.M.E. and Bernold, L.E. (2013). Factors influencing construction technology adoption. In Stephen Kajewski, A.P.K.M.a.P.K.H. (Ed.), *19th CIB World Building Congress, Brisbane 2013: Construction and Society.* Queensland University of Technology, Brisbane

Sheikhshoaei, F., Naghshineh, N., Alidousti, S. and Nakhoda, M. (2018). Design of a digital library maturity model (DLMM). *The Electronic Library*, 36(4): 607–619

Solanke, B.H. and Fapohunda, J.A. (2015). *Impacts of E-commerce on Construction Materials Procurement for Sustainable Construction.* World Congress on Sustainable Technologies (WCST), London, 14–16 December

Solis, B. (2016). *The Six Stages of Digital Transformation Maturity.* Altimeter Cognizant. Available at: www.prophet.com/2016/04/the-six-stages-of-digital-transformation/#:~: text=To%20make%20this%20more%20actionable,%2C%20Technology%20Integra tion%2C%20Digital%20Literacy [accessed 6–8–2020]

Tang, W., Qiang, M., Duffield, C.F., Young, D.M. and Lu, Y. (2009). Enhancing total quality management by partnering in construction. *Journal of Professional Issues in Engineering Education and Practice*, 135(4): 129–141

Teece, D.J. (2007). Explicating dynamic capabilities: The nature and micro foundations of sustainable enterprise performance. *Strategic Management Journal*, 28(8): 1319–1350

Teece, D.J. (2010). Business models, business strategy and innovation. *Long Range Planning*, 43: 172–194

Teece, D.J. and Pisano, G. (1994). The dynamic capabilities of firms: An introduction. *Industrial and Corporate Change*, 3(3): 537–556

Teece, D.J., Pisano, G. and Shuen, A. (1997). Dynamic capabilities and strategic management. *Strategic Management Journal*, 18(7): 509–533

Telkom. (2016). Digital readiness assessment – why a digital readiness assessment is important for your business. Available at: www.telkom.co.za/bigbusiness [accessed 19–11–2019]

Teo, H.H., Wei, K.K. and Benbasat, I. (2003). Predicting intention to adopt interorganizational linkages: An institutional perspective. *MIS Quarterly*, 27(1): 19–49

Thong, J. (1999). An integrated model of information systems adoption in small businesses. *Journal of Management Information Systems*, 15(4): 27–31

Tidd, J. (2001). Innovation management in context: Environment, organization and performance. *International Journal of Management Reviews*, 3: 169–183

Tornatzky, L. and Fleischer, M. (1990). *The Process of Technology Innovation*. Lexington Books, Lexington, MA

Tsuja, P.Y. and Mariño, J.O. (2013). The influence of the environment on organizational innovation in service companies in Peru. *Review of Business Management*, 15: 582–600

Vaduva-Sahhanoglu, A., Calbureanu-Popescu, M.X. and Smid, S. (2016). Automated and robotic construction-a solution for the social challenges of the construction sector. *Revista de stiinte politice*, 50: 1–11

Vähä, P., Heikkilä, T. and Kilpeläinen, P. (2013). Extending automation of building construction – survey on potential sensor technologies and robotic applications. *Automation in Construction*, 36: 168–178

Valdez-de-Leon, O. (2016). A digital maturity model for telecommunications service providers. *Technology Innovation Management Review*, 6(8): 19–32

Vivares, J.A., Sarache, W. and Hurtado, J.E. (2018). A maturity assessment model for manufacturing systems. *Journal of Manufacturing Technology Management*, 29(5): 746–767

Vollmer, C. and Egor, M. (2014). Five rules for strategic partnerships in a digital world. *Strategy + Business*. Available at: www.strategy-business.com/blog/Five-Rules-for-Strategic-Partnerships-in-a-Digital-World?gko=ebbc1 [accessed 6–8–2020]

Wang, Y. (2000). Coordination issues in Chinese large building projects. *Journal of Management in Engineering*, 16(6): 54–61

Wikström, K., Artto, K., Kujala, J. and Söderlund, J. (2010). Business models in project business. *International Journal of Project Management*, 28(8): 832–841

Wong, P.S.P. and Cheung, S.O. (2005). Structural equation model of trust and partnering success. *Journal of Management in Engineering*, 21(2): 70–80

Yeung, J., Chan, A. and Chan, D. (2007). The definition of alliancing in construction as a Wittgenstein family-resemblance concept. *International Journal of Project Management*, 25(3): 219–231

Yoo, Y., Boland Jr, R.J., Lyytinen, K. and Majchrzak, A. (2012). Organizing for innovation in the digitized world. *Organization Science*, 23(5): 1398–1408

Yoo, Y., Henfridsson, O. and Lyytinen, K. (2010). The new organizing logic of digital innovation: An agenda for information systems research. *Information Systems Research*, 21(4): 724–735

Zhang, X., Majid, S. and Foo, S. (2011). The contribution of environmental scanning to organizational performance. *Singapore Journal of Library and Information Management*, 40: 65–88

Zhou, K.Z. and Wu, F. (2010). Technology capability, strategic flexibility and product innovation. *Strategic Management Journal*, 3(5–6): 547–561

Zott, C. and Amit, R. (2008). The fit between product market strategy and business model: Implications for firm performance. *Strategic Management Journal*, 29: 1–26

9 Exploring the conceptualised digitalisation capability maturity model in a developing country through expert opinion

Introduction

This chapter presents the result of the assessment of the suitability and applicability of the dimensions and sub-attributes of the conceptualised DCMM in the construction industry using South Africa as a case study. A Delphi approach was adopted in the assessment. The chapter gives a clear description of Delphi approach and the steps involved in conducting a Delphi study. The chapter further presents that all six dimensions and their sub-attributes were adjudged as suitable to construction digitalisation by experts.

Exploring the applicability of the conceptualised DCMM

The conceptual model proposed in Chapter 8 revealed six major capability areas (dimension) wherein construction organisations need to attain significant maturity to be digitally transformed. These are technology, people, process, strategy, digital partnering, and environment. The Delphi approach was employed to explore these dimensions' suitability and applicability in the construction industry in a developing country. This was done following the suggestion in past studies that exploratory study can further help strengthen the outcome and acceptance of a maturity model (De Bruin et al., 2005). The South African construction industry was used as a case study since the country is a developing country with the potential of rapid digital transformation in diverse sectors of the country (Dall'Omo, 2017).

The Delphi was developed as a forecasting tool that can help solve complex issues using expert opinion (Skulmoski et al., 2007). This approach has been described as a structured communication and consensus attaining method conducted among a group of experts on a problem that is considered complex (Chan et al., 2001; Gohdes & Crews, 2004). The Delphi requires experts within a field to give answers to non-leading unambiguous statements with the sole purpose of achieving consensus (Holey et al., 2007). These answers can forecast future occurrences and create solutions to envisaged problems (Agumba & Musonda, 2013). The approach could be subjective and intuitive at the same time (Alomari et al., 2018). In affirmation, Aigbavboa (2013) noted that the approach is based on a structured survey and uses intuitively available information from the Delphi experts. The approach also gives qualitative and quantitative results and has explorative and predictive capabilities.

The Delphi involves an iterative process where consensus is often reached through rounds of feedback from experts' opinion and judgement on the subject under review (Hallowell & Gambatese, 2010). The method's strength lies in its rigorous query of experts through an iterative process that provides initial feedback, analyses of feedback, and presentation of analysed feedback to the experts for further evaluation (Agumba & Musonda, 2013). Similarly, the anonymity in the process allows for a more reliable response by the experts without pressure (Aigbavboa, 2013).

The Delphi design and execution

Expert panel selection and size

The criteria for selecting an expert panel for a Delphi are crucial and must be given considerable attention as any shortcoming in this area might flaw the whole Delphi process. This area has been criticised as one of the Delphi shortcomings since in most cases the reliability of the experts selected for the study is difficult to determine (Gupta & Clarke, 1996; Keeney *et al.*, 2001). On a broad spectrum, for an expert to be selected for the Delphi process they must have the knowledge and experience in the study area, they must have the capacity and willingness to undertake the survey, they must have sufficient time to be part of the Delphi process since the process is iterative and can be time-consuming. They must possess effective communication skills.

Hallowell and Gambatese (2010) suggested that an unbiased method is necessary for the selection of experts. In selecting these experts Linstone and Turoff (1975) warn against taking the part of "least resistance" by selecting a group of individuals with like-minded people, or a group of cosy friends which invariably negates the purpose of the entire process and might lead to poor Delphi outcome. In literature, several criteria exist for the selection of experts in a Delphi process. While some researchers have adopted the process of setting specific requirements for experts' selection (Chan *et al.*, 2001; Manoliadis *et al.*, 2006), others have proposed a flexible point system (Hallowell & Gambatese, 2010; Hallowell *et al.*, 2011). For example, Rogers and Lopez (2002) suggested that for an individual to be termed expert, they must meet at least two out of the following criteria. These are, "*authorship, conference presenter, member or chair of the committee, employed in practice with five years of experience, and employed as a faculty member at an institute of higher learning*". While these have been adopted in some Delphi studies, Hallowell and Gambatese (2010) argued that the authorship and conference presentations requirements are considerably vague and open to interpretation. Like in the case of this current study where the concept of digitalisation is fairly new within developing countries (South Africa inclusive) and the adoption of digital technology is at its early stage (Aghimien *et al.*, 2018; Osunsanmi *et al.*, 2018), getting authors of journal publications and books is almost unlikely. Therefore, the aspect of authorship and conference presentation was not considered. In comparing the risk perception of construction safety professionals in the USA, using the flexible points system, Alomari *et al.* (2018) identified nine criteria and expected at least 50% overall value for a participant to be selected. These criteria include,

> Holding a PhD with an emphasis on safety research, book or book chapter editing, a faculty member at an accredited institution of higher learning, above five years' experience in the construction, membership of a construction safety committee, invited to present at a conference with a focus on construction safety, above three conference papers on construction safety, above three journal articles on construction safety, and professional registration.

From the past studies, it is evident that researchers adopt the most suitable approach to the subject under review. To this end, this study draws from Alomari *et al.* (2018) procedure by expecting at least 50% overall value for a participant to be selected based on all the following criteria for expert selection. These criteria are:

(1) Experts must have extensive professional experience within the South African construction industry with above five years of working experience (Alomari *et al.*, 2018; Chan *et al.*, 2001; Hallowell & Gambatese, 2010).
(2) An expert must be a faculty member at an accredited higher institution or a current active employee of a construction organisation with a considerable number of projects being handled (Chan *et al.*, 2001; Hallowell & Gambatese, 2010; Rogers & Lopez, 2002).
(3) An expert must be a member of a professional body such as the South African Council for the Architectural Profession, South African Council of Quantity Surveying Profession, Engineering Council of South Africa, and the South African Council for Project and Construction Management Professions.
(4) An expert must possess a higher education degree with a minimum of bachelor's in a construction-related field.

In terms of the sample size for a Delphi, there is no consensus in the body of literature regarding the ideal number for a Delphi panel. While some researchers believe that the larger the size of the panel, the more reliable the result will be (Murphy *et al.*, 1998), others believe that there is no significant relationship between the size of the panel and the accuracy and effectiveness of the result generated (Boje & Murnighan, 1982; Brockhoff, 1975). For this reason, the literature has revealed a diverse sample size in the Delphi conducted. For example, Rowe *et al.* (1991) suggested that the size of a Delphi panel can range from three to 80 panellists, but Hallowell and Gambatese (2010) and Skulmoski *et al.* (2007) noted that a sample size of between ten and 18 expert members is typical. Hallowell and Gambatese (2010) suggested a minimum of eight panellists is ideal for the process. Ameyaw *et al.* (2016) assessed 67 papers in construction engineering management which adopted the Delphi approach and observed that the number of experts ranged from three to 93. However, it was noted that most researchers tend to employ a panel size of between eight and 20 experts.

Xu *et al.* (2010) used 34 experts in a two-round Delphi process to develop a fuzzy risk allocation model for public–private partnership projects in China. Also, in the study of Tymvios and Gambatese (2016), 17 experts participated in developing ways of designing construction workers' safety in the USA. Alomari *et al.* (2018) sought the opinion of 48 experts with only 21 completing the process in an attempt to compare the risk perception of construction safety professionals in the USA. Similarly, due to the insufficiency in theory and expertise in smart city development, Rana *et al.* (2018) employed the view of eight experts to unearth the barriers to the development of smart cities in India. Eyiah-Botwe (2017) explored some factors relating to stakeholder management in the Ghanaian construction industry using ten experts. Similarly, Tengan and Aigbavboa (2018) used 11 experts in validating the factors influencing monitoring and evaluation in the Ghanaian construction industry.

For this study, a total of 32 panellists were invited to participate in the Delphi through emails and telephone conversations. The invited panellists were given a summary of what the Delphi process is all about and its application to the subject under review. Out of these invited panellists, 21 indicated their interest and availability to participate in the study. As such, 21 designed questionnaires were sent out for the first round along with a form on the background information for the experts to fill. However, out of the 21 experts who participated from the start of the survey, only 13 completed the process. This number of respondents is believed to be well in the range of acceptable number for a Delphi process, as seen in previous studies.

Based on the aforementioned criteria set for expert selection and size, Table 9.1 shows that all the 13 experts who participated in the study has above 50% criteria set for selection.

Table 9.1 Experts' qualification assessment

S/n	Experts Select Criteria	Panellists												
		1	2	3	4	5	6	7	8	9	10	11	12	13
1	Above 5 years working experience in the South Africa Construction Industry	x	x	x	x	x	x	x	x	x	x	x	x	x
2	Expert must be a faculty member at an accredited higher institution or a current active employee of a construction organisation	x	x	x	x	x	x	x	x	x	x	x	x	x
3	Professional Membership	–	x	x	x	x	x	x	–	x	–	–	–	x
4	Higher education degree (Bachelor, Masters, PhD)	x	x	x	x	x	x	x	x	x	x	x	x	x
	Overall percentage (%) for all criteria	75	100	100	100	100	100	100	75	100	75	75	75	100

Conducting the Delphi iterations

In Delphi, the term iteration is used to describe the feedback process, which is done in a series of rounds (Hsu & Sanford, 2007). Several studies have adopted several numbers of rounds in achieving consensus, as no specific guidance in respect to the ideal number of rounds for a Delphi process exists in the body of literature. However, a consensus has been observed to be reached in rounds two or three of most Delphi studies in construction engineering and management research (Ameyaw et al., 2016). While Dalkey et al. (1970) noted that the result from a Delphi process tends to be more accurate after two rounds, Hasson et al. (2000) warned that a Delphi process that is more than three rounds might incur problems such as fatigue of the experts, time, cost, and increased attrition rates.

For this study, two iterations were used to attain a consensus. The Delphi questionnaire for round one was developed based on the conceptual model developed and was sent out to the respondents electronically. The questionnaire was designed to address the objectives of the Delphi. The questionnaire adopted for the second round was designed based on the feedback obtained from the first round. The second round was conducted to allow experts to review and comment on digitalisation attributes of construction organisations. It was also geared towards giving feedback to the experts on analysing the data gathered in the first round of the exercise and requesting their affirmation and comments on attributes and variables that did not receive significant consideration in the first round. The questions in the round were asked in a closed question form. Analyses of the feedback from the second round revealed that consensus was reached for about 80% of the variables asked. This is in line with suggestion of Green et al. (1999) that consensus of at least 80% of the assessed variables is adequate for a conclusion to be drawn. Each round of the process took a month to complete.

Computation of data from Delphi survey and determining consensus

In gathering data from experts using the Delphi questionnaire, the experts were asked to rate the main dimensions of attaining digital transformations using an importance range from zero to 100% with at interval range of ten as shown in Table 9.2. Similarly, the experts were asked to rate the significance of the sub-attributes needed for attaining digital transformations on a ten-point Likert scale ranging from "no significance" to "very high significance as shown in Table 9.3.

Table 9.2 Importance scale

0–10%	11–20%	21–30%	31–40%	41–50%	51–60%	61–70%	71–80%	81–90%	91–100%
1	2	3	4	5	6	7	8	9	10
							X		

Table 9.3 Significance scale

No significance		Low significance		Average significance		High significance		Very high significance	
1	2	3	4	5	6	7	8	9	10
							X		

The responses were transferred to a Microsoft Excel spreadsheet and Statistical Package for Social Sciences (SPSS), and the group median for each section was computed. Considering the small sample of the Delphi, the group median was deemed appropriate due to its ability to eliminate bias and consider outlier responses (Aigbavboa, 2013).

Chan *et al.* (2001) have noted that the Delphi's primary purpose is to achieve consensus among experts regarding an issue under review. Likewise, Skulmoski *et al.* (2007) have opined that there is a linear relationship between the method of data collection and analysis and the reliability of the survey outcome. The literature on Delphi studies has shown that researchers have employed diverse approaches to attain consensus and make sense out of data gathered (Aigbavboa, 2013; Ameyaw *et al.*, 2016; Holey *et al.*, 2007; Rayens & Hahn, 2000). Ameyaw *et al.* (2016) noted that three major techniques had been adopted in most studies within the construction engineering and management field. These are Deviation, Kendall's coefficient of concordance (W), and Chi-square (X^2). However, out of these three techniques, the Deviation techniques seem to be the most used, followed by Kendall's W. The least favoured is the Chi-square technique. However, studies combining these approaches have also emerged in recent times (Ojo & Ogunsemi, 2019).

While the standard deviation approach along with the mean value for each variable is common (Hallowell & Gambatese, 2010; Hallowell *et al.*, 2011; Hsu & Sandford, 2007; Rayens & Hahn, 2000; Holey *et al.*, 2007), the interquartile deviation (IQD) along with the median has been suggested by other studies as the most reliable deviation approach towards attaining consensus (Aigbavboa, 2013; Tengan & Aigbavboa, 2018; Hasson *et al.*, 2000). The IQD is calculated using the absolute value of the difference between the 75th and 25th percentiles. The percentile is described as *"the value below, which a certain per cent of observation fall"* (Aigbavboa, 2013). The first quartile is the 25th percentile, while the second quartile, which is also the median, is the 50th percentile, and the 75th percentile is the third quartile. The deviation between the first and the third quartiles gives the IQD needed in achieving consensus. The smaller the IQD, the higher the degree of consensus for each variable. According to Raskin (1994), consensus can be reached when each variable achieves an IQD score less than or equal to one. Rayens and Hahn (2000) went further to suggest that an IQD of at most one should be attained for more than 60% of the variables under assessment.

Other studies have also favoured the use of Kendall's W in analysing data gathered in a Delphi (Hallowell *et al.*, 2011; Hon *et al.*, 2012; Ojo & Ogunsemi, 2019;

Tymvios & Gambatese, 2016; Xia *et al.*, 2009). Hon *et al.* (2012) noted that Kendall's W value shows the degree of agreement between experts by considering the variations between rankings of the mean of the different variables. A concordance coefficient of one implies 100% consensus. Thus, the value of concordance coefficient is expected to increase for each round of the Delphi process since in most cases experts are expected to confirm their response from the previous rounds in line with the group responses. Ameyaw *et al.* (2016) stated that most of the published Delphi studies within the construction engineering and management field had achieved consensus with Kendall's W value ranging between 0.234 and 0.600. The analysis conducted using SPSS gives Kendall's W value and the X^2 value with an associated degree of freedom (Df). It was suggested that the X^2 value should be adopted when the number of variables to be evaluated is larger than seven (Siegel & Castellan, 1988) as in the case of this current study, and the consensus is termed as achieved when computed X^2 value is larger than the critical X^2 value derived from a statistical table.

To achieve robustness in the analysis of data gathered from experts and attain genuine consensus, this study followed the path of Ojo and Ogunsemi (2019) by combining together the IQD, Kendall's W, and X^2 techniques using SPSS. In summary, the IQD of 60% of the assessed variable must be a score that is less than or equal to one, Kendall's W must be closer to one, and the computed X^2 value must be larger than the critical X^2 value derived from a statistical table for consensus to be achieved.

Delphi outcome

A total of 32 experts were identified and invited to participate in this study, out of which only 21 indicated their interest to participate. Based on this acceptance, 21 questionnaires were sent out for the first round and a form on the background information for the experts to fill with 13 responses obtained. These 13 experts all met the criteria earlier set for experts' selection. The result in Table 9.4 reveals that more PhD holders participated in the Delphi process, followed by those with bachelors and master's degree. Out of the 13 experts, six were construction managers, three were engineers, three were quantity surveyors, and one was an architect. Eight out of the 13 panellists were members of a professional body in the South African construction industry while seven are not. In terms of years of working experience, all the panellists were within the criteria of above five years' working experience with eleven of them having above ten years and only two having between five to ten years' working experience in the construction industry. Also, seven of these panellists currently work within an academic institution while six are practising within a construction organisation that is handling a considerable number of projects within the country. These panellists are in three provinces in the country with more responses from experts in Gauteng (9) followed by the Free state with three responses, and Mpumalanga with one response.

Table 9.4 Background information of panellists

Category	Classification	Frequency	Percentage
Highest academic qualification	Bachelors	4	30.8
	Masters	2	15.4
	PhD	7	53.8
	Total	13	100
Profession	Architect	1	7.7
	Engineer	3	23.1
	Construction Manager	6	46.2
	Quantity Surveyors	3	23.1
	Total	13	100.0
Member of a professional body	Yes	8	61.5
	No	5	38.5
	Total	13	100.0
Years of experience in the construction industry	5 to 10 years	2	15.4
	Above 10 years	11	84.6
	Total	13	100.0
Current employer	Academic institution	7	53.8
	Construction organisation	6	46.2
	Total	13	100.0
Province	Gauteng	9	69.2
	Free State	3	23.1
	Mpumalanga	1	7.7
	Total	13	100.0

Main dimensions for construction digitalisation

Following Aigbavboa's (2013) suggestion that using group median is most appropriate for a Delphi analysis due to its ability to eliminate bias and take into consideration outlier responses, the result in Table 9.5 shows the median for the first-round responses. The table also shows the IQD of each main dimension and the z-value and significant p-value derived from Mann-Whitney U-Test conducted. Since the experts were drawn from both the academic environment and the construction organisations, it became necessary to ascertain if there is any significant difference in the experts' opinion from both environments. The choice of adopting Mann-Whitney U-Test as against other alternatives such as T-test was premised because it compares medians of the groups and converts the scores on the continuous variable to ranks and determines the significant difference between both groups. This contrasts with what is obtainable in T-test, where the mean values of the two groups are compared (Pallant, 2005). It is important to note that when the derived p-value is lower than the predetermined significance value of 0.05, it means that there is a significant difference in the median value of the two groups (Pallant, 2005). In this case, it means that there is a significant difference in the median value of the experts from academic institutions and construction organisations.

Table 9.5 Round 1 results of main dimensions of digitalisation capability maturity

Main dimensions	Median	IQD	Mean	SD	Mann-Whitney	
					Z	Sig.
People	10	1.00	9.46	0.776	−0.657	0.511
Process	10	2.00	9.15	1.068	−1.487	0.137
Strategies	10	1.00	9.15	1.281	−1.105	0.269
Technological	9	2.00	9.08	0.954	−0.233	0.816
Digital partnering	8	5.00	7.69	2.394	−0.220	0.826
Environment	8	3.00	6.92	2.660	−0.516	0.606
Cronbach alpha					0.654	
Kendall's W					0.335	
X^2					21.765	
X^2 - Critical values from statistical table (p=0.05)					11.070	
Df					5	
Sig.					0.001	

However, the reverse is the case if the derived p-value is greater than the predetermined significance value of 0.05. This means that there is no significant difference in the view of experts from both environments.

Table 9.5 shows that in the first round, the experts considered all the six main dimensions to be important to the digitalisation of construction organisations in South Africa. However, only people and strategies had IQD of 1.00, implying that consensus was not attained for the remaining four dimensions. Mann-Whitney U-Test shows that at a confidence level of 95% there is no difference in the view of experts from both groups as a p-value of above 0.05 was derived for all assessed dimensions. A Cronbach alpha value of 0.654 was derived, which shows considerable reliability of the instrument since reliability is said to be attained when the alpha value is from 0.6 and above (Moser & Kalton, 1999). Kendall's W gave a value of 0.335, which implies that this expert level of consensus is low at this stage. However, Ameyaw et al. (2016) have noted that most Delphi studies that have adopted Kendall's W have achieved consensus with a concordance coefficient value ranging from 0.234 to 0.600. An X^2 value of 21.765 at a Df of 5 and significant value of 0.001 was derived. This X^2 is greater than the critical X^2 value (11.070) derived from the statistical table.

The result in Table 9.6 reveals that the group median from the first round was retained by the experts, while IQD of 0.00 was derived for four dimensions and 1.00 for two others. Based on the criteria set for consensus to be achieved at IQD of 1.00 and below, it can be deduced that there was consensus among the experts for all the assessed dimensions. Similarly, Mann-Whitney U-Test revealed that at a confidence level of 95%, there is no difference in the view of experts from both groups as p-value of above 0.05 was derived for all assess dimensions. A Cronbach alpha value of 0.656 was derived for this round, implying that the instrument used was reliable. Kendall's W of 0.678 was derived, thus confirming 68% consensus among the experts in rating these dimensions. Comparing this figure with the

Table 9.6 Round 2 results of main dimensions of digitalisation capability maturity

Main dimensions	Median	IQD	Mean	SD	Mann-Whitney	
					Z	Sig.
People	10	0.00	9.85	0.376	−0.657	0.511
Strategies	10	0.00	9.77	0.439	−1.105	0.269
Process	10	0.00	9.77	0.439	−1.487	0.137
Technological	9	1.00	9.15	0.801	−0.233	0.816
Digital partnering	8	1.00	8.54	0.776	−0.220	0.826
Environment	8	0.00	8.23	0.725	−0.516	0.606
Cronbach alpha					0.656	
Kendall's W					0.678	
X^2					44.098	
X^2 - *Critical values from statistical table (p=0.05)*					11.070	
Df					5	
Sig.					0.000	

0.335 derived from the first round, the Kendall's W value further confirms Hon *et al.* (2012) submission that the value of the concordance coefficient is expected to increase for each round of the Delphi process since in most cases experts are expected to confirm their response from the previous rounds in line with the group responses. The computed X^2 value of 44.098 derived is also higher than the critical X^2 value of 11.070 derived from the statistical table. Based on the result, it can be deduced that at this stage, significant consensus was achieved among the experts as regards the importance of the six dimensions assessed. The experts considered these dimensions important to the digital transformation of construction organisations with the level of importance ranging from 80% to 100%.

Sub-attributes for construction digitalisation

Technology capability

Table 9.7 shows that although experts rated the 15 identified sub-attributes of technology as significant (between high and very high) in the first round, there was no consensus for most of these sub-attributes. The IQD for nine of the sub-attributes were above 1.0 while the overall Kendall's W was 0.162, which is low. The computed X^2 value of 24.919 was slightly higher than the critical X^2 value of 23.685 derived from the statistical table. Mann-Whitney U-test revealed no statistically significant difference in these experts' view as a *p*-value of above 0.05 was derived for all attributes. Since opportunity was given to the experts to add any new sub-attribute they think can help with digitalisation and that is not included on the survey, one of the experts noted the need for "technology transfer". This was then included in the round two survey for the panellists to rate its level of significance.

For the second round, a Cronbach alpha value of 0.820 was derived, which shows that the instrument used is highly reliable. Analysis of the response revealed

Table 9.7 Round 1 results of the technology capability

Technology	Median	IQD	Mean	SD	Mann–Whitney	
					Z	Sig.
Technology planning	8	1.00	8.23	1.536	–0.220	0.826
Emerging technologies optimisation	9	2.00	8.69	1.109	–1.110	0.267
Big data analytics	9	1.00	7.92	2.499	–0.733	0.464
Additive manufacturing	9	2.00	7.77	2.619	–0.950	0.342
Cloud computing	9	1.00	8.38	1.387	–0.453	0.651
Standard IT systems	9	1.25	9.08	0.996	–0.172	0.863
Design modelling (BIM usage)	9	1.00	9.00	1.414	–1.681	0.093
Design simulation and virtualisation	9	2.00	8.46	2.537	–0.379	0.705
Robotics and automation in construction	8	3.00	7.38	2.725	–0.579	0.563
Digital media marketing	8	2.00	6.75	2.701	–0.500	0.617
Cybersecurity in organisations	9	3.00	8.23	1.878	–1.692	0.091
Technology governance	9	2.00	8.77	1.013	–0.375	0.708
Continuous technology upgrades	9	1.00	9.15	0.689	–0.791	0.429
Flexible technology usage	9	2.00	9.08	1.038	–0.685	0.494
The synergy of technology with other organisation's resources	8	1.00	8.08	2.326	–0.446	0.656
Technology transfer	–	–	10.00	–	–	–
Group mean			8.44			
Cronbach alpha					0.778	
Kendall's W					0.162	
X^2					24.919	
X^2 - Critical values from statistical table (p=0.05)					23.685	
Df					14	
Sig.					0.035	

that the experts rated 13 of the assessed sub-attributes of technology as very high and two as just high. However, there seems to be some disparity among the experts from academic institutions and those within the industry in rating sub-attributes such as "design modelling", and "cybersecurity in organisations". *p*-Values of 0.009 and 0.034 were derived for both sub-attributes, which are lower than the 0.05

threshold. Although there is a disparity in their rating, these two sub-attributes are still considered very significant as a group median of 9 was derived for both variables (See Table 9.8). The table also shows that consensus was achieved for all the sub-attributes as an IQD of 0.00 and 1.00 was derived, and Kendall's W was 0.403 which falls within the range of value adopted for consensus in other past studies (see Ameyaw *et al.*, 2016) was derived. Further look at the table shows that the computed X^2 value of 66.449 was higher than the critical X^2 value of 23.685 derived from the statistical table, implying a robust consensus among the experts for the identified sub-attributes for technology capability. It is imperative to note that the sub-attribute "technology transfer" earlier suggested in the first round by one of the experts was considered by other experts to be part of "technology

Table 9.8 Round 2 results of the technology capability

Technology	Median	IQD	Mean	SD	Mann–Whitney	
					Z	Sig.
Flexible technology usage	9	1.00	9.08	0.760	−0.384	0.701
Continuous technology upgrades	9	0.00	9.08	0.494	−0.581	0.561
Design modelling (BIM usage)	9	1.00	9.08	0.760	−2.611	0.009**
Standard IT systems	9	0.00	9.08	0.641	−0.493	0.622
Design simulation and virtualisation	9	1.00	9.00	1.080	−1.658	0.097
The synergy of technology with other organisation's resources	9	0.00	8.92	0.641	−0.411	0.681
Technology governance	9	1.00	8.92	0.669	−0.458	0.647
Big data analytics	9	0.00	8.85	0.555	−0.880	0.379
Additive manufacturing	9	0.00	8.77	0.725	−0.699	0.484
Cybersecurity in organisations	9	0.00	8.77	1.013	−2.125	0.034**
Cloud computing	9	0.00	8.77	0.725	−0.262	0.793
Emerging technologies optimisation	9	1.00	8.69	0.855	−1.452	0.147
Technology planning and transfer	8	0.00	8.15	0.801	−0.327	0.744
Digital media marketing	8	0.00	8.15	0.801	−1.389	0.165
Robotics and automation in construction	8	1.00	8.08	0.862	−1.026	0.305
Group Mean			8.76			
Cronbach alpha					0.820	
Kendall's W					0.403	
X^2					66.449	
X^2 - Critical values from statistical table (p=0.05)					23.685	
Df					14	
Sig.					0.000	

** Significant at p < 0.05

planning" which was already given on the survey. Therefore, the sub-attribute was modified to "technology planning and transfer" as shown in Table 9.8.

People capability

The role of people in the digital transformation of any industry has been emphasised in past studies. The analysis of the first round of the Delphi revealed that most of this dimension's sub-attributes are deemed very significant to the digitalisation of construction organisations (see Table 9.9). Mann-Whitney U-Test revealed no significant difference in the views of the two groups of experts in the rating of these sub-attributes as a significant *p*-value of above 0.05 was derived. No consensus

Table 9.9 Round 1 results of the people capability

People	Median	IQD	Mean	SD	Mann–Whitney	
					Z	Sig.
Digital technical know-how of personnel	10	2.00	9.15	1.405	−1.486	0.137
Digital knowledge management by the organisation	8	2.00	8.38	1.850	−0.306	0.760
Reskilling of workforce	10	1.00	9.46	0.877	−0.880	0.379
Continuous learning of personnel	10	1.00	9.38	1.193	−0.262	0.793
Top management support	10	0.00	9.85	0.376	−0.114	0.909
Digital culture within an organisation	9	2.00	9.00	1.080	−0.076	0.939
Organisation's positive change attitude	9	2.00	8.85	1.214	−0.375	0.708
Digital empowerment of personnel	9	2.00	8.62	1.502	−0.075	0.940
Personnel's innovativeness	9	3.00	8.23	1.922	−0.441	0.659
Attracting the right digital talent	8	2.00	7.77	1.833	−1.107	0.268
Retaining the right digital talent	8	1.00	8.08	1.935	−1.026	0.305
Group mean			8.80			
Cronbach alpha					0.904	
Kendall's W					0.345	
X^2					44.790	
X^2 - Critical values from statistical table (p=0.05)					18.307	
Df					10	
Sig.					0.000	

Table 9.10 Round 2 results of the people capability

People	Median	IQD	Mean	SD	Mann–Whitney	
					Z	Sig.
Top management support	10	0.00	9.85	0.376	−0.114	0.909
Continuous learning of personnel	10	0.00	9.77	0.439	−0.488	0.626
Reskilling of workforce	10	0.00	9.77	0.599	−1.363	0.173
Digital technical know-how of personnel	10	0.00	9.77	0.599	−0.227	0.820
Digital culture within an organisation	9	1.00	9.15	0.801	−0.762	0.446
Organisation's positive change attitude	9	1.00	9.08	0.760	−1.152	0.249
Digital empowerment of personnel	9	1.00	9.00	0.913	−0.459	0.647
Personnel's innovativeness	9	0.00	8.85	1.068	−0.078	0.938
Retaining the right digital talent	8	1.00	8.23	0.927	−0.082	0.935
Attracting the right digital talent	8	0.00	8.08	0.760	−0.880	0.379
Digital knowledge management by the organisation	8	0.00	8.08	0.494	−0.581	0.561
Group mean			9.06			
Cronbach alpha					0.856	
Kendall's W					0.686	
X^2					89.167	
X^2 – *Critical values from statistical table* (p=0.05)					18.307	
Df					10	
Sig.					0.000	

was reached at this stage as most of the variables had an IQD of between 0.00 and 3.00, and Kendall's W value of 0.345 was also derived. However, the computed X^2 value of 44.790, which is higher than the critical X^2 value of 18.307 was derived. Although this is high enough for consensus, it is not enough to justify consensus at this stage due to the low values derived from the IQD and Kendall's W.

The result in Table 9.10 shows that a Cronbach alpha value of 0.856 was derived for the second round of the Delphi process. The table also shows that experts considered eight out of the 11 assessed sub-attributes to have a very high significance. Mann-Whitney U-Test revealed that there is no significant difference in the views of the two groups of experts in the rating of these sub-attributes as a p-value of above 0.05 was derived. The IQD revealed a strong consensus of between 0.00 and 1.00, while Kendall's W value of 0.686, close to 1.0 was also derived. To further affirm the attainment of consensus for these sub-attributes, the computed X^2 value of 89.167 was higher than the critical X^2 value of 18.307 obtainable in statistical tables.

Process capability

The Project Management Body of Knowledge (PMBOK, 2017) gave the ten project management knowledge areas essential for the smooth delivery of any project (construction inclusive). These knowledge areas are considered as key operations/

process need for the project delivery. In the context of this study, digitalising these processes can lead to the attainment of better service delivery. Aside from these ten areas, the aspect of change management, digitalised planning, and construction processes have also been reiterated in past studies. Thus, this study assessed 13 sub-attributes needed for the digital transformation of construction organisations regarding the process of project delivery. The result in Table 9.11 revealed that while all other sub-attributes were considered significant, communication management is believed to have very high significance to the digitalisation of the construction process. Mann-Whitney U-Test revealed no significant difference in the views of the two groups of experts in the rating of these sub-attributes as a *p*-value of above 0.05 was derived. No consensus was reached at this stage as most of the variables had an IQD of between 2.00 and 3.00 and Kendall's W value of 0.070. The computed X^2 value of 10.067 was also lower than the critical X^2 value of 19.675.

Table 9.11 Round 1 results of the process capability

Process	Median	IQD	Mean	SD	Mann–Whitney	
					Z	Sig.
Digitalised planning process	8	3.00	8.38	1.758	−0.303	0.762
Automated construction process	8	3.00	7.69	2.562	−0.294	0.769
Digitalised schedule management	8	3.00	8.15	1.625	−0.957	0.339
Digitalised resource management	8	3.00	7.85	1.819	−0.295	0.768
Management of integrated service	8	3.00	8.50	1.382	−0.346	0.729
Value delivery and quality management	8	2.00	7.77	1.833	−0.293	0.769
Digitalised stakeholder management	8	3.00	7.00	2.517	−1.096	0.273
Digitalised scope management	8	3.00	8.08	2.178	−0.518	0.604
Digitalised cost management process	8	3.00	7.31	2.562	−1.174	0.240
Digitalised procurement process	8	3.00	8.31	0.947	−1.795	0.073
Digital risk management	8	3.00	8.00	1.683	−0.672	0.502
Change management	8	2.00	8.00	1.225	−1.769	0.077
Communication management	9	2.00	8.69	1.109	−0.148	0.882
Group mean			7.98			
Cronbach alpha					0.931	
Kendall's W					0.070	
X^2					10.067	
X^2 - Critical values from statistical table (p=0.05)					19.675	
Df					12	
Sig.					0.610	

Table 9.12 Round 2 results of the process capability

Process	Median	IQD	Mean	SD	Mann–Whitney	
					Z	Sig
Communication management	9	1.00	8.62	0.650	−1.431	0.153
Digital risk management	8	0.00	8.08	1.038	−0.962	0.336
Digitalised scope management	8	1.00	8.08	1.188	−1.094	0.274
Management of integrated service	8	0.00	8.08	0.862	−0.861	0.389
Change management	8	1.00	8.00	0.913	−1.146	0.252
Value delivery and quality management	8	0.00	8.00	0.816	−1.230	0.219
Digitalised schedule management	8	0.00	7.92	1.188	−1.636	0.102
Automated construction process	8	0.00	7.92	1.115	−1.919	0.055
Digitalised planning process	8	0.00	7.92	1.188	−0.234	0.815
Digitalised resource management	8	0.00	7.85	1.144	−1.305	0.192
Digitalised cost management process	8	0.00	7.77	1.092	−1.923	0.054
Digitalised procurement process	8	0.00	7.77	0.440	−0.488	0.626
Digitalised stakeholder management	8	0.00	7.69	1.377	−1.387	0.165
Group mean			7.98			
Cronbach alpha					0.926	
Kendall's W					0.097	
X^2					15.144	
X^2 - *Critical values from statistical table (p=0.05)*					19.675	
Df					12	
Sig.					0.284	

The result in Table 9.12 shows a Cronbach alpha value of 0.926, which implies high reliability of the survey instrument used. Communication management was considered the most significant sub-attribute for this dimension as a group median of 9 was derived. The remaining 12 sub-attributes have a group median of 8 which implies high significance. Furthermore, Mann-Whitney U-Test revealed no significant difference in the views of the two groups of experts in the rating of these sub-attributes as a significant *p*-value of above 0.05 was derived. Although Kendall's W gave a value of 0.097 which is considered very low and the computed X^2 value of 15.144 which is also lower than the critical X^2 value of 19.675 was derived, the IQD of all the sub-attributes revealed a robust consensus. IQD of between 0.00 was derived for ten sub-attributes and 1.00 for only three. In this instance, the IQD was used to justify consensus, and this was further corroborated by the Mann-Whitney test, which revealed significant agreement in the view of the two groups of experts in rating these sub-attributes.

Strategy capability

The strategy adopted by an organisation in the attainment of competitive advantage through better service delivery and achieving digital transformation is crucial. The result in Table 9.13 reveals that while seven of the 14 sub-attributes of this

Table 9.13 Round 1 results of the strategy capability

Strategy	Median	IQD	Mean	SD	Mann–Whitney	
					Z	Sig.
Research and development	10	0.00	9.62	0.870	–0.581	0.561
Digital technology selection	8	2.00	7.69	2.136	–0.444	0.657
Forecasting client's need	8	2.00	8.23	1.589	–0.597	0.551
Choice of market selection	8	2.00	7.31	2.097	–1.094	0.274
Project and service boundary	8	2.00	7.15	2.340	–0.581	0.561
Client's feedback evaluation	9	2.00	9.00	0.816	–0.681	0.496
Finance and investment in digital technology	9	1.00	8.92	1.498	–1.679	0.093
Decision-making procedure	8	1.00	8.38	1.261	–0.440	0.660
Digitalised performance measurement	9	1.00	8.08	1.801	–0.592	0.554
Error-proofing approaches	8	1.00	7.23	2.488	–0.382	0.703
Information and communication management	9	2.00	8.85	1.144	–0.745	0.457
Digital alignment of business	9	2.00	8.46	1.664	–0.222	0.824
Digital risk culture within the organisation	8	3.00	8.31	1.601	0.000	1.000
Open innovation within the organisation	9	2.00	8.92	1.256	–0.917	0.359
Group mean			8.30			
Cronbach alpha					0.824	
Kendall's W					0.289	
X^2					48.786	
X^2 – Critical values from statistical table (p=0.05)					22.362	
Df					13	
Sig.					0.000	

dimension were believed to have very high significance (group median of 9 and 10) at the first round of the Delphi process. Mann-Whitney U-test revealed no significant difference in the views of the two groups of experts in the rating of these sub-attributes as a *p*-value of above 0.05 was derived. No consensus was reached at this stage as most of the variables had an IQD of between 0.00 and 3.00, and Kendall's W value of 0.289 was also derived. However, the computed X^2 value of 43.786 which is higher than the critical X^2 value of 22.362 was derived. While this is high enough for consensus, it is not enough to justify consensus for all sub-attributes at this stage due to the low values derived from the IQD and Kendall's W.

The result in Table 9.14 reveals a Cronbach alpha value of 0.867 for the second round of the Delphi process, and this implies that the survey instrument was highly reliable. A total of seven sub-attributes were deemed to be very significant with a group median of 9 and 10. At the same time, the remaining seven were considered to be significant with a group median of 8. Mann-Whitney U-Test revealed that at 95% confidence level, there is a statistically significant difference in the views of

Table 9.14 Round 2 results of the strategy capability

Strategy	Median	IQD	Mean	SD	Mann–Whitney	
					Z	Sig.
Research and development	10	0.00	9.77	0.439	−0.488	0.626
Finance and investment in digital technology	9	1.00	9.15	0.899	−2.016	0.044**
Open innovation within the organisation	9	1.00	9.08	0.954	−0.834	0.404
Digital alignment of business	9	1.00	9.08	0.760	−0.307	0.759
Information and communication management	9	1.00	8.92	0.760	−1.075	0.282
Client's feedback evaluation	9	0.00	8.92	0.641	−1.315	0.189
Digitalised performance measurement	9	1.00	8.69	0.480	−1.336	0.181
Digital risk culture within the organisation	8	1.00	8.38	0.961	−1.441	0.150
Forecasting client's need	8	0.00	8.23	0.927	−2.207	0.027**
Decision-making procedure	8	0.00	8.15	0.801	−1.389	0.165
Choice of market selection	8	0.00	7.92	1.188	−1.636	0.102
Project and service boundary	8	1.00	7.77	1.301	−1.281	0.200
Digital technology selection	8	0.00	7.69	0.947	−1.923	0.054
Error-proofing approaches	8	0.00	7.62	2.063	−1.312	0.189
Group mean			8.53			
Cronbach alpha					0.867	
Kendall's W					0.530	
X^2					89.590	
X^2- Critical values from statistical table (p=0.05)					22.362	
Df					13	
Sig.					0.000	

the two groups of experts in the rating of two sub-attributes (finance and investment in digital technology and forecasting client's need) as a *p*-value of below 0.05 was derived. However, a significant agreement exists among both groups in rating the other 12 sub-attributes with a *p*-value of above 0.05 derived. The IQD revealed a strong consensus of between 0.00 and 1.00 for all 14 sub-attributes, while Kendall's *W* value of 0.530 was also derived. In confirmation of the attainment of consensus for these sub-attributes, the computed X^2 value of 89.590 was higher than the critical X^2 value of 22.362 obtainable in statistical tables.

Digital partnering capability

The result in Table 9.15 revealed that three out of the eight sub-attributes assessed under the digital partnering dimension were considered to have a very high significance in the first round of the Delphi survey. The remaining five were also rated as

Table 9.15 Round 1 results of the digital partnering capability

Digital partnering	Median	IQD	Mean	SD	Mann–Whitney	
					Z	Sig.
Digital Partner selection	9	3.00	8.38	1.660	–0.515	0.606
Long-term commitment to digital partnering	8	2.00	7.62	2.142	–0.369	0.712
Top management support	9	2.00	8.85	1.214	–0.225	0.822
Conflict resolution in digital partnering	8	1.00	7.23	2.088	–0.700	0.484
Trust in digital partners	9	0.00	8.62	1.502	–0.156	0.876
Common goal in the partnering process	8	1.00	8.08	1.656	–0.741	0.459
Digital partnering coordination	8	0.00	8.00	1.472	–0.782	0.434
Structured digital workshop and training	8	3.00	7.62	1.981	–0.292	0.770
Group mean			8.05			
Cronbach alpha					0.744	
Kendall's W					0.186	
X^2					16.933	
X^2 - Critical values from statistical table (p=0.05)					14.067	
Df					7	
Sig.					0.018	

being significant with a group median of 8. Mann-Whitney U-Test revealed no significant difference in the views of the two groups of experts in the rating of these sub-attributes as a p-value of above 0.05 was derived. No consensus was reached at this stage as most of the variables had an IQD of between 0.00 and 3.00, and Kendall's W value of 0.186 was also derived. However, the computed X^2 value of 16.933 was slightly higher than the critical X^2 value of 14.067 derived from the statistical table.

Analysis of the second round of the Delphi response in Table 9.16 revealed a Cronbach alpha value of 0.641, which implies that the survey instrument was reliable. Trust in digital partners, top management support, and digital partner selection were significant to digital partnering as a group median of 9 was derived for all three sub-attributes. The remaining five sub-attributes were considered significant as a group median of 8 was derived for them. Mann-Whitney U-Test revealed that at 95% confidence level, there is a statistically significant difference in the views of the two groups of experts in the rating of "structured digital workshop and training" as a p-value of 0.036 was derived. However, a significant agreement exists among both groups in rating the other seven sub-attributes with a p-value of above 0.05 derived. The IQD revealed a strong consensus of between 0.00 and 1.00 for all the sub-attributes, while Kendall's W value of 0.494 derived. In affirmation of the attainment of consensus for these sub-attributes, the computed X^2 value of 44.981 was higher than the critical X^2 value of 14.067 obtainable in the statistical table.

Table 9.16 Round 2 results of the digital partnering capabilities

Digital partnering capabilities	Median	IQD	Mean	SD	Mann–Whitney	
					Z	Sig
Trust in digital partners	9	0.00	9.00	0.577	−0.964	0.335
Top management support	9	0.00	8.92	0.641	−1.233	0.218
Digital partner selection	9	0.00	8.92	0.641	−0.493	0.622
Common goal in the partnering process	8	1.00	8.31	0.751	−1.559	0.119
Long-term commitment to digital partnering	8	1.00	8.31	0.751	−1.230	0.219
Structured digital workshop and training	8	0.00	8.23	0.725	−2.098	0.036**
Digital partnering coordination	8	0.00	8.15	0.689	−1.739	0.082
Conflict resolution in digital partnering	8	0.00	7.77	0.927	−1.642	0.101
Group mean			8.45			
Cronbach alpha					0.641	
Kendall's W					0.494	
X^2					44.981	
X^2 – Critical values from statistical table (p=0.05)					14.067	
Df					7	
Sig.					0.000	

Environment capability

The result in Table 9.17 revealed that two sub-attributes were noted to have a very high significance in the first round. Mann-Whitney U-Test also revealed that there is no significant difference in the views of the two groups of experts in the rating of these sub-attributes as a *p*-value of above 0.05 was derived. At this stage, no consensus was reached as most of the variables had an IQD of between 1.00 and 3.00 and Kendall's W value of 0.418. However, the computed X^2 value of 32.638 was higher than the critical X^2 value of 12.592 derived from the statistical table.

Analysis of the second-round response in Table 9.18 revealed a Cronbach alpha value of 0.817, which implies that the survey instrument used was reliable. Most significant of the sub-attributes in this dimension is the client's pressure with a group median of 9. The remaining six sub-attributes were considered significant as a group median of 7 and 8 were derived for them. Mann-Whitney U-Test also revealed no significant difference in the views of the two groups of experts in the rating of these sub-attributes as a *p*-value of above 0.05 was derived. The IQD revealed a strong consensus of between 0.00 and 1.00 for all seven sub-attributes, while Kendall's W value of 0.637 was also derived. In affirming the attainment of consensus for these sub-attributes, the computed X^2 value of 49.678 was higher than the critical X^2 value of 12.592 obtainable in the statistical table.

Table 9.17 Round 1 results of the environment capability

Environment	Median	IQD	Mean	SD	Mann–Whitney	
					Z	Sig.
Suppliers' pressure	7	1.00	7.38	1.261	−0.220	0.826
Clients' pressure	9	2.00	8.69	1.182	−0.300	0.764
Competitors influence	9	2.00	8.62	1.261	−0.370	0.712
Government legislation	8	2.00	7.38	2.501	−0.657	0.511
Industry rules and regulations	8	2.00	7.38	2.567	−0.515	0.606
Environmental impact assessment	7	3.00	6.00	2.517	−0.075	0.940
Market scope	8	2.00	7.00	1.633	−0.980	0.327
Group mean			7.49			
Cronbach alpha					0.807	
Kendall's W					0.418	
χ^2					32.638	
χ^2 - Critical values from statistical table (p=0.05)					12.592	
Df					6	
Sig.					0.000	

Table 9.18 Round 2 results of the environment capability

Environment	Median	IQD	Mean	SD	Mann–Whitney	
					Z	Sig
Clients' pressure	9	1.00	8.92	0.760	−0.307	0.759
Competitors influence	8	1.00	8.85	0.689	−1.502	0.133
Industry rules and regulations	8	1.0	8.38	0.768	−0.789	0.430
Government legislation	8	1.0	8.15	0.899	−0.456	0.648
Market scope	8	0.0	7.92	0.494	−1.646	0.100
Environmental impact assessment	7	0.0	7.08	0.760	−1.056	0.291
Suppliers' pressure	7	0.00	6.92	0.862	−1.644	0.100
Group mean			8.03			
Cronbach alpha					0.817	
Kendall's W					0.637	
χ^2					49.678	
χ^2 - Critical values from statistical table (p=0.05)					12.592	
Df					6	
Sig.					0.000	

Overall view of the structural dimension of digitalisation capability of construction organisations

Using the group mean derived for each dimension at the second round of the Delphi, Figure 9.1 shows the level of significance attributed to each dimension of the conceptualised digitalisation capability maturity model by construction experts in South Africa. The figure shows that the most significant dimension wherein maturity is required for digital transfromation of construction organisations is

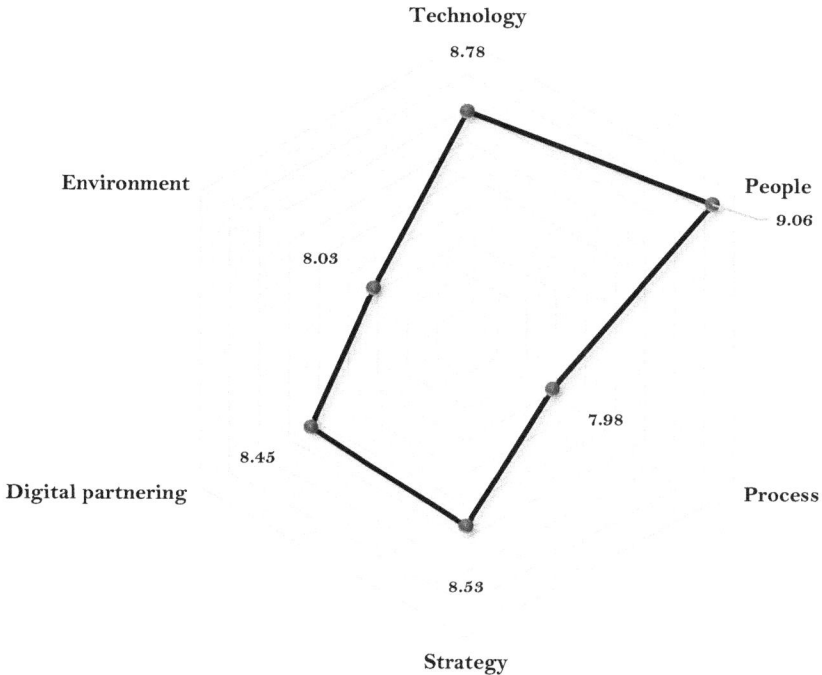

Figure 9.1 Overall significance of the digitalisation capability maturity dimensions

"people". This is followed by technology, strategy, digital partnering, and environment, with the least being process. Therefore, irrespective of the process of service delivery, with maturity in the people and their ability to use acquired technologies and implement the developed strategy, attaining digitalisation might be more of a dream than a reality for construction organisations.

Main findings

At the end of the second round, all six dimensions of the conceptualised model were considered significant as they all had a median of between 8 and 10. Judging from their overall mean derived from the average mean of all the sub-attributes, people, technology, strategies, digital partnering, and environment all have reasonably high significance. This finding further affirms Kane *et al.'s* (2015) submission that the strength of digitalisation is not in the individual technologies being adopted, but rather it lies in the company's ability to integrate them into their daily activities or business models to transform their businesses and operations. This integration is evident in their capability to adopt the right strategy, invest in digital technology, manage their human resources, form a strong digital partnership with other organisations with complementary digital resources, and be able to transform the pressure from the organisation to be digitalised.

In terms of the variables needed for measuring these dimensions, Delphi revealed that all the identified sub-attributes in the conceptualised model are significant in the South African context. For the people capability, eight sub-attributes were deemed to have very high significance by the experts. These attributes include support from top management, continuous learning, reskilling of the workforce, digital technical know-how, digital culture, positive change attitude within organisations, digital empowerment of personnel, and personnel's innovativeness. The importance of management support in successful technology adoption has been reiterated in past studies (Gill *et al.*, 2016; Quinto *et al.*, 2018). In the same vein, this finding aligns with the submissions of Bennis (2013), Boström and Celik (2017), and Kane *et al.* (2015) on the need to upgrade digital skills within organisations. This can be achieved through an awareness of positive change attitude in the organisations. The finding also aligns with Sheikhshoaei *et al.*'s (2018) submission that promoting a digital culture well as digitally empowering of employees will go a long way in the digital transformation of an organisation.

The strategy dimension revealed the top variable to be R&D. This further confirms Teece's (2007) submission that for an organisation seeking a good dynamic capability, it must sense opportunities that lie within its immediate environment. This can only be achieved through proper investment in R&D. As for the process within the organisation, management of communication was rated as the highest capability variable for digitalisation. This can be a result of the fact that there is a need for a hitch-free communication system for construction service to be delivered effectively. However, although this variable is an important project management knowledge area (PMBOK, 2017), its mention in existing models reviewed is minimal. More attention has been given to change management, managing integrated services, value delivery management, and automated processes (Boström & Celik, 2017; Gerbert *et al.*, 2016; Newman, 2017), which are also considered significant process variables by the Delphi experts in this current case study.

Trust in digital partners, top management support, and digital partner selection were very significant in the digital partnering dimension. This finding is in line with previous studies that have shown that the success of a partnering process is mainly dependent on partner selected (Cheng & Li, 2002; Kadefors, 2004; Nystrom, 2005). It is also in tandem with the submissions of Das and Teng (2001), and Wong and Cheung (2005) who noted that trust is crucial for effective partnering relationship. Creating a trustworthy environment where there is belief in the partners' digital competence and their integrity as it relates to fulfilling the partnering agreements is highly important (Nystrom, 2005). Similarly, this finding aligns with past studies that have revealed that top management support is important for partnering to occur (Black *et al.*, 2000; Cheng & Li, 2004; Conley & Gregory, 1999). The environment dimension revealed that client's pressure is the most significant variable needed to promote digitalisation in an organisation. This submission is in tandem with the findings from past studies that have shown that adopting an innovative idea can either be promoted or thwarted by the client's involvement (Gupta *et al.*, 2013; Gutierrez *et al.*, 2015; Quinton *et al.*, 2018).

Summary

This chapter explored the suitability and applicability of the dimensions and sub-attributes of the conceptualised DCMM in a developing economy's construction industry. Using a Delphi approach, the study concluded that all these six dimensions and their sub-attributes were adjudged as suitable for construction digitalisation by experts.

References

Aghimien, D.O., Aigbavboa, C.O. and Oke, A.E. (2018). Digitalisation for effective construction project delivery in South Africa. *Contemporary Construction Conference: Dynamic and Innovative Built Environment (CCC2018)*, Coventry, 5–6 July, pp. 3–10

Agumba, J. and Musonda, I. (2013). Experience of using Delphi method in construction health and safety research. *Seventh International Conference on Construction in the 21st Century (CITC-VII)*, Bangkok, Thailand, December 19–21

Aigbavboa, C. (2013). *An Integrated Beneficiary Centred Satisfaction Model for Publicly Funded Housing Schemes in South Africa*. A PhD thesis submitted to the Post Graduate School of Engineering Management, University of Johannesburg, Johannesburg

Alomari, K.A., Gambatese, J.A. and Tymvios, N. (2018). Risk perception comparison among construction safety professionals: Delphi perspective. *Journal of Construction Engineering & Management*, 144(12): 1–12

Ameyaw, E.E., Hu, Y., Shan, M., Chan, A.P.C. and Le, Y. (2016). Application of Delphi method in construction engineering and management research: A quantitative perspective. *Journal of Civil Engineering and Management*, 22(8): 991–1000

Bennis, W. (2013). Leadership in a digital world: Embracing transparency and adaptive capacity. *Management Information Systems Quarterly*, 37(2): 635–636

Black, C., Akintoye, A. and Fitzgerald, E. (2000). An analysis of success factors and benefits of partnering in construction. *The International Journal of Project Management*, 18(6): 423–432

Boje, D.M. and Murnighan, J.K. (1982). Group confidence pressures decisions. *Management Science*, 28(1): 1187–1196

Boström, E. and Celik, O.C. (2017). *Towards a Maturity Model for Digital Strategizing – A Qualitative Study of How an Organisation Can Analyse and Assess Their Digital Business Strategy*. IT Management Master thesis submitted to the Department of Informatics, UMEA Universitet, Sweden

Brockhoff, K. (1975). The performance of forecasting groups in computer dialogue and face-to-face discussion. In *The Delphi Method: Techniques and Applications*. Addison-Wesley, Reading, MA, pp. 291–321

Chan, A.P.C., Yung, E.H.K., Lam, P.T.I., Tam, C.M. and Cheung, S.O. (2001). Application of Delphi method in selection of procurement systems for construction projects. *Construction Management and Economics*, 19(7): 699–718

Cheng, E. and Li, H. (2002). Construction partnering process and associated critical success factors: Quantitative investigation. *Journal of Management in Engineering*, 18(4): 194–202

Cheng, E. and Li, H. (2004). Development of a practical model of partnering for construction projects. *Journal of Construction Engineering & Management*, 130(6): 790–798

Conley, M.A. and Gregory, R.A. (1999). Partnering on small construction projects. *Journal of Construction Engineering & Management*, 125(5): 320–324

Dalkey, N., Brown, B. and Cochran, S. (1970). Use of self-ratings to improve group estimates. *Technological Forecasting*, 1(3): 283–291

Dall'Omo, S. (2017). Driving African development through smarter technology. *African Digitalisation Maturity Report*: 1–45

Das, T.K. and Teng, B.S. (2001). Trust, control and risk in strategic alliances: An integrated framework. *Organisation Studies*, 22(4): 253–285

De Bruin, T., Freeze, R., Kaulkarni, U. and Rosemann, M. (2005). Understanding the main phases of developing a maturity assessment model. In Campbell, B., Underwood, J. and Bunker, D. (Eds.), *Australasian Conference on Information Systems (ACIS)*, Australia, New South Wales, Sydney, November 30–December 2

Eyiah-Botwe, E. (2017). *Development of a Sustainable Stakeholder Management Framework for Construction Projects in Ghana*. A PhD thesis submitted to the Post Graduate School of Engineering Management, University of Johannesburg, Johannesburg

Gerbert, P., Castagnino, S., Rothballer, C., Renz, A. and Filitz, R. (2016). *Digital Engineering and Construction: The Transformative Power of Building Information Modelling*. The Bolton Consulting Group

Gill, M., VanBoskirk, S., Evans, P.F., Nail, J., Causey, A. and Glazer, L. (2016). The digital maturity model 4.0. Benchmarks: Digital business transformation playbook. Available at: www.forrester.com

Gohdes, S.L.W. and Crews, B.T. (2004). The Delphi technique: A research strategy for career and technical education. *Journal of Career and Technical Education*, 20(2): 55–67

Green, B., Jones, M., Hughes, D. and Williams, A. (1999). Applying the Delphi technique in a study of GP's information requirement. *Health and Social Care in the Community*, 17(3): 198–205

Gupta, P., Seetharaman, A. and Raj, J. (2013). The usage and adoption of cloud computing by small and medium businesses. *International Journal of Information Management*, 3(5): 861–874

Gupta, U.G. and Clarke, R.E. (1996). Theory and applications of the Delphi technique: A bibliography (1975–1994). *Technological Forecasting and Social Change*, 53(2): 185–211

Gutierrez, A., Boukrami, A. and Lumsden, R. (2015). Technological, organisational and environmental factors influencing managers' decision to adopt cloud computing in the UK. *Journal of Enterprise Information Management*, 28(6): 788–807

Hallowell, M.R., Esmaeili, B. and Chinowsky, P. (2011). Safety risk interactions among highway construction work tasks. *Construction Management and Economics*, 29(4): 417–429

Hallowell, M.R. and Gambatese, J. (2010). Qualitative research: Application of the Delphi method to CEM research. *Journal of Construction Engineering & Management*, 136(1): 99–107

Hasson, F., Keeney, S. and McKenna, H. (2000). Research guidelines for the Delphi survey technique. *Journal of Advanced Nursing*, 32(4): 1008–1015

Holey, E.A., Feeley, J.L., Dixon, J. and Whittaker, V.J. (2007). An exploration of the use of simple statistics to measure consensus and stability in Delphi studies. *BMC Medical Research Methodology*, 7(52): 1–10

Hon, C.K.H., Chan, A.P.C. and Yam, M.C.H. (2012). Empirical study to investigate the difficulties of implementing safety practices in the repair and maintenance sector in Hong Kong. *Journal of Construction Engineering and Management*, 138(7): 877–884

Hsu, C. and Sandford, B.A. (2007). The Delphi technique: Making sense of consensus. *Practical Assessment Research and Evaluation*, 12(10): 1–8

Kadefors, A. (2004). Trust in project relationships – inside the black box. *The International Journal of Project Management*, 22(3): 175–182

Kane, G.C., Palmer, D., Phillips, A.N., Kiron, D. and Buckley, N. (2015). *Strategy, Not Technology, Drives Digital Transformation*. MIT Sloan Management Review and Deloitte University Press, Westlake, TX

Keeney, S., Hasson, F. and McKenna, H.P. (2001). A critical review of the Delphi technique as a research methodology for nursing. *International Journal of Nursing Studies*, 38(2): 195–200

Linstone, H.A. and Turoff, M. (1975). *The Delphi Method: Techniques and Applications*. Addison Wesley, Reading, MA, p. 620

Manoliadis, O.G., Tsolas, I. and Nakou, A. (2006). Sustainable construction and drivers of change in Greece: A Delphi study. *Construction Management and Economics*, 24(2): 113–120

Moser, C.A. and Kalton, G. (1999). *Survey Methods in Social Investigation*, 2nd edition. Gower Publishing Company, Aldershot

Murphy, M.K., Black, N., Lamping, D.L., McKee, C.M., Sanderson, C.F.B., Askham, J. and Marteau, T. (1998). Consensus development methods and their use in clinical guideline development. *Health Technology Assessment*, 2(3): 1–88

Newman, M. (2017). Digital maturity model (DMM): A blueprint for digital transformation (TM Forum White Paper). Available at: www.tmforum.org

Nystrom, J. (2005). The definition of partnering as a Wittgenstein family resemblance concept. *Construction Management and Economics*, 23: 473–481

Ojo, L. and Ogunsemi, D. (2019), Critical drivers (CDs) of value management adoption in the Nigerian construction industry. *Journal of Engineering, Design and Technology*, 17(1): 250–264

Osunsanmi, T.O., Aigbavboa, C.O. and Oke, A.E. (2018). Construction 4.0: The future of South Africa construction industry. *World Academy of Science, Engineering and Technology International Journal of Civil and Environmental Engineering*, 12(3): 206–212

Pallant, J. (2005). *SPSS Survival Manual: A Step-by-Step Guide to Data Analysis Using SPSS for Windows (Version 12)*, 2nd edition. Allen and Unwin, Crows Nest NSW2065, Australia

Project Management Body of Knowledge PMBOK. (2017). *A Guide to Project Management Body of Knowledge*, 6th edition. Project Management Institute, Philadelphia, PA

Quinton, S., Canhoto, A., Molinillo, S., Pera, R. and Budhathoki, T. (2018). Conceptualising a digital orientation: Antecedents of supporting SME performance in the digital economy. *Journal of Strategic Marketing*, 26(50): 427–439

Rana, N.P., Luthra, S., Kumar, S., Mangla, S.K., Islam, R., Roderick, S. and Dwivedi, Y.K (2018). Barriers to the development of smart cities in Indian context. *Information Systems Frontiers*: 1–23

Raskin, M.S. (1994). The Delphi study in field instruction revisited: Expert consensus on issues and research priorities. *Journal of Social Work Education*, 30: 75–89

Rayens, M.K. and Hahn, E.J. (2000). Building consensus using the policy Delphi method. *Policy Politics Nursing Practice*, 1(2): 308–315

Rogers, M.R. and Lopez, E.C. (2002). Identifying critical cross-cultural school psychology competencies. *The Journal of Social Psychology*, 40(2): 115–141

Rowe, G., Wright, G. and Bolger, F. (1991). Delphi – a re-evaluation of research and theory. *Technological Forecasting and Social Change*, 39: 238–251

Sheikhshoaei, F., Naghshineh, N., Alidousti, S. and Nakhoda, M. (2018). Design of a digital library maturity model (DLMM). *The Electronic Library*, 36(4): 607–619

Siegel, S. and Castellan, N.J. (1988). *Nonparametric Statistics for the Behavioral Sciences*, 2nd edition. McGraw-Hill, New York

Skulmoski, J.G., Hartman, T.F. and Krahn, J. (2007). The Delphi method for graduate research. *Journal of Information Technology Education*, 6: 1–21

Teece, D. (2007). Explicating dynamic capabilities: The nature and micro foundations of sustainable enterprise performance. *Strategic Management Journal*, 28(8): 1319–1350

Tengan, C. and Aigbavboa, C. (2018). Validating factors influencing monitoring and evaluation in the Ghanaian construction industry: A Delphi study approach. *International Journal of Construction Management*, 10(2): 1–12

Tymvios, N. and Gambatese, J.A. (2016). Direction for generating interest for design for construction worker safety – a Delphi study. *Journal of Construction Engineering & Management*, 142(8): 1–11

Wong, P.S.P. and Cheung, S.O. (2005). Structural equation model of trust and partnering success. *Journal of Management in Engineering*, 21(2): 70–80

Xia, B., Chan, A.P.C. and Yeung, J.F.Y. (2009). Identification of key competences of design-builders in the construction market of the People's Republic of China (PRC). *Construction Management and Economics*, 27(11): 1141–1152

Xu, Y., Chan, A.P.C., Yeung, J.F.Y., Chan, D.W.M., Wang, S. and Ke, K. (2010). Developing a risk assessment model for PPP projects in China: A fuzzy synthetic evaluation approach. *Journal of Automation in Construction*, 19(7): 929–943

Part V

Conclusion

Part V of this book has one chapter, which is the concluding aspect of the book.

10 Conclusion and recommendations

Introduction

The general objective was to present a digitalisation capability maturity model (DCMM) that can assess the capability maturity of construction organisations, particularly in developing countries. By exploring existing maturity models in different domains, the study conceptualised the DCMM for construction organisations. The study examined the suitability of the capability areas and their sub-attributes in a developing country – South Africa. Therefore, this last chapter concludes the book by highlighting the significant conclusions drawn from the different areas covered in the study. The chapter also outlines the theoretical and practical contributions of the conceptualised DCMM.

Conclusion

The construction industry is unique in its product delivery, and its contribution to the socio-economic development of any nation is significant. The industry is responsible for delivering infrastructure, employment opportunities and economic contribution to nations' GDP. Exploring the body of knowledge, it is concluded that while the industry in developed countries continues to thrive, the story is not the same for those in developing countries. These construction industries struggle as a result of the diverse challenges existing within their environment. Issues of the poor image of the industry, poor health and safety, lack of synergy between industry participants, unethical practices, resistance to change and innovation, lack of human, financial, and technological resources, unfavourable government policies, economic conditions and poverty among others have bedevilled the construction industry in these developing countries. These issues have left most construction organisations bewildered and unable to deliver projects within set targets. The result of this problem is the perpetual state of poor project delivery with failure to complete projects within agreed budget and time, failure to meet agreed specification leading to poor quality and client's dissatisfaction, poor health, and safety delivery in the course of construction and unsustainable construction projects that have characterised the industry.

To solve these poor performance problems, advancement in technology offers some "silver lining" for the construction industry. The fourth industrial revolution (4IR) offers a plethora of ubiquitous technologies that can help solve the industry's age-long problem. The 4IR which involves the use of diverse emerging and disruptive digital technologies in the discharge of more productive activities offers the construction industries in developing countries the opportunity to leapfrog and measure up with their counterparts in the developed countries. To achieve this, digitalisation which is a critical driver of the 4IR, is essential. By digitalising construction activities in the different phases of project delivery, construction organisations stand a chance to improve their performance and attain the satisfaction of clients and better competitive advantage. Digital technologies such as IoT, big data analytics, 3D printing, BIM, sensors, UAVs, AI, machine learning, AR, VR, blockchain, and digital twin can help digitally transform the industry.

Evidence has pointed to the fact that despite the clamour for the adoption of these beneficial digital technologies, the construction industry in both developed and developing countries is still slow to embrace digitalisation. The situation is worse for those in developing countries who are deterred by their unique problems of unavailability of technical expertise, finance, resistance to change, knowledge and awareness, policies and legislations, lack of standardisation among others. The result of this is the lack of innovativeness evident in the construction industry of these countries. While technologies such as BIM and cloud computing might be gaining prominence in some of these developing countries, more still need to be done to improve digital technologies in the quest for better project delivery and a more digitally transformed industry.

It is pertinent to note that the adoption of digital technologies is not a "silver bullet" for the total alienation of the problems of the construction industry in developing countries but an option that promises significant benefits. Careful planning and implementation are needed for the adoption of these technologies to yield the expected result. Knowing what to adopt, when to adopt, and how to adopt is essential for the success of the transformation process. Carefully assessing the risk that might come with the adoption of digital technologies is also necessary at the onset to understand the appropriate measures to be taken when they occur. Risk issues surrounding finance, loss of productivity, the uncertainty of the outcomes of the adopted digital tools, adoption of technologies in silos, security, job losses, legal, psychological issues, and the likes must be carefully evaluated with mitigating measures put in place to cushion their effect on the organisation should they occur.

Aside from understanding the risk and putting measures in place, construction organisations need to assess their digitalisation capability maturity as they embark on their digital transformation journey. Knowing where they are and where they ought to be in terms of their digital capability is essential to the success of their digital transformation. Regrettably, no capability maturity model for construction digitalisation exists within the body of literature. The available digital maturity models have emanated from other domains such as business, education, manufacturing, and telecommunication. Just like these different domains, the construction industry is business-oriented and delivers finished products to its consumers.

Therefore, some of the variables identified in these existing digital maturity models from other domains can be explored within the construction industry. Based on this understanding, four major capability areas were identified from existing models: technology, people, process, and strategy. However, considering the slow adoption of digital technology in the construction industry, it is believed that an ideal capability maturity model should include digital partnering wherein construction organisations can partner with other organisations within and outside the industry to share digital resources with a long-term commitment of attaining digital transformation. Also, since construction organisations do not operate in isolation, considering their capability to handle pressure from the environment and converting such pressure into a positive drive that will help in their digitalisation is worth considering. Based on this backdrop, the study achieved its set objective of delivering a conceptualised DCMM. The conceptualised model was further explored within the construction industry of a developing country (South Africa). The capability areas and sub-attributes were adjudged to be suitable and applicable to construction experts' construction industry through a Delphi study.

Contribution to knowledge

Theoretical contribution

The developed DCMM provides an excellent platform for future studies to build upon since there is no digitalisation capability maturity model tailored to address construction organisations' maturity to be digitalised. Also, while some organisations might be willing to adopt digital technologies, how to go about it might be a problem. For those who have already adopted some digital technologies, sustaining them for continuous competitive advantage and better service delivery is a challenge. Thus, it is evident that this study contributes theoretically to the body of knowledge in terms of the road map needed for the digitalisation of construction organisations.

The study lends its voice to the digitalisation discourse by theorising that a construction organisation that will attain significant digital transformation, provide better service delivery for its client, and in the process obtain a competitive edge over its competitors, must strive to achieve maturity in six digital capabilities areas, namely technology, people, process, strategy, digital partnering, and environment. The improvement in the maturity level of these capabilities will see construction organisations improve their digital uptake, digital culture, digital readiness, become digitally transformed, offer better project delivery to cost, time, conformance to quality, and social sustainability, increase their productivity, and improve their overall project performance.

Practical contribution

The practical contribution of this research can be seen in the conceptualised DCMM. Unlike the past, where construction organisations implement without guidance, the developed model can now serve as a road map towards attaining digital transformation. Construction organisations can utilise the conceptualised model to assess their current digital capability maturity and understand where they need to improve. Through proper self-assessment and continuous learning and improvement, improved digitalisation of construction organisations can be attained. This is because the model recognises the importance of digitalising vital areas of the construction process. Based on the submission of the study, construction organisations can now see the importance of having a well-defined organisational and managerial process, good positioning of their strategic assets, and a clear path. This is essential as they form the bases of construction organisations' dynamic capabilities to be digitally transformed.

The DCMM reveals that construction organisation in the quest for digitalisation must attain maturity in technology by acquiring and using digital technologies in its day-to-day function. Upskilling and equipping the organisation's human resources and providing support for the development of people within the organisation to handle digital technologies requires attention. Furthermore, construction organisations must strive to attain maturity in the processes involved in the delivery of construction projects right from the planning and design phase through to the operation and demolition phase. The strategy adopted by the organisation to attract and retain clients is also a crucial capability area wherein maturity must be attained. Support digitalisation research and investing in digital technology, along with other strategic systems need to be given adequate attention. Maturity is also required in digital partnering as it is easier for digital transformation to be achieved when organisations take the digitalisation journey in the company of other organisations with complementary resources and experience. Finally, construction organisations must be matured enough to handle the pressure from the environment. These pressures must serve as a driving force for these organisations in attaining digital transformation.

Recommendations

If the construction industry in developing countries is to grow and actively perform their role of nation-building and socio-economic development, adequate consideration must be given to technology advancement. Construction organisations must be ready to jettison the old construction project delivery method and embrace a more digitalised approach that promises effective and efficient project delivery.

It is imperative for construction organisations seeking digital transformation to put measures to handle possible risks associated with financial investment, loss of productivity, cyber-security, information overload, potential system failure, job loss, silo implementation of digital technologies, legal issues, violation of compliances among others. These are some of the identified risks of digitalising the construction processes noted in the study, which can be mitigated using the right risk mitigation strategy.

In digitalising, construction organisations must understand that there must be a synergy between the different capability areas provided in the DCMM.

Considering one aspect and ignoring the other might lead to a hampered digital transformation process. While considering optimising the use of emerging and disruptive technologies in the construction delivery process, consideration must be given to the right strategy to be adopted and the human resources needed to drive the organisation's strategy. Drawing from other organisations (within and outside the construction industry) that have the required digital knowledge and resources through effective digital partnering must also be given thorough consideration. The pressure coming from the environment must also be considered as this might harm the whole transformation process if not properly handled.

It is noteworthy to state that the conceptualised DCMM in this study provides a broad view of the capability areas and their sub-attributes needed for the assessment of the digitalisation capability maturity of construction organisations, particularly in developing countries. Therefore, practitioners and researchers can further test the suitability and applicability of the conceptualised model in the context of their country or region.

Index

Note: Page numbers in *italic* indicate a figure and page numbers in **bold** indicate a table on the corresponding page.